NORTH AMERICAN
F-86 SABRE

1947 onwards (all day-fighter variants)

ZENITH PRESS

Haynes
®

Published in North America in 2011 by Zenith Press,
an imprint of MBI Publishing Company,
400 1st Avenue North, Suite 300, Minneapolis,
MN 55401 USA, by arrangement with
Haynes Publishing.

Zenith Press titles are also available at discounts in
bulk quantity for industrial or sales-promotional use.
For details write to Special Sales Manager at
MBI Publishing Company, 400 1st Avenue North,
Minneapolis, MN 55401 USA.

To find out more about our books, join us online at
www.zenithpress.com or www.qbookshop.com.

ISBN-13: 978-0-7603-4292-3

Printed in the USA by Odcombe Press LP,
1299 Bridgestone Parkway, La Vergne, TN 37086

Zenith Press

NORTH AMERICAN
F-86 SABRE

1947 onwards (all day-fighter variants)

Owners' Workshop Manual

Haynes

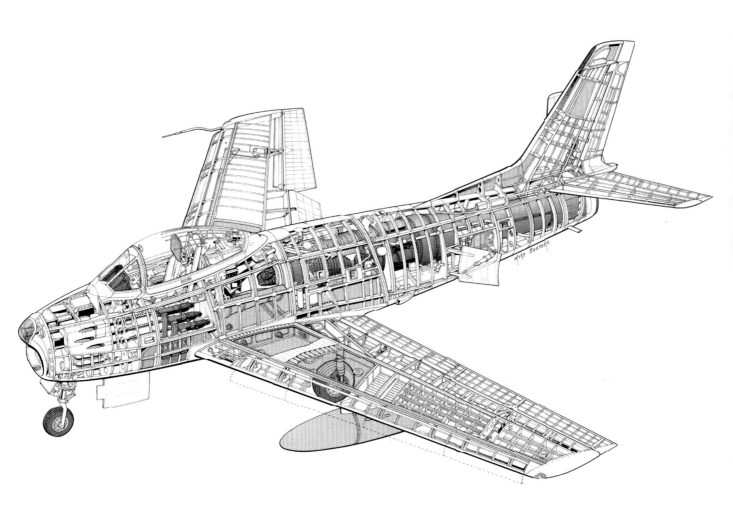

An insight into owning, flying, and maintaining
the USAF's legendary Cold War jet fighter

Mark Linney

Contents

OPPOSITE John Dibbs' fine air-to-air study of 48-178 taken in Cambridgeshire skies in 2010. *(John Dibbs 2010)*

Acknowledgements

ABOVE Staff from the Duxford based Aircraft Restoration Company are a superb team with many different skillsets and who all share the same passion for this F-86A, 48-178. For them all it has been a true labour of love.

This book would not have been possible without the help and inspiration of many others.

I was first introduced to the Sabre by Mark and Ray Hanna of the Old Flying Machine Company. Both of these fine pilots shared their incredible knowledge and experience with me: they are still very much missed.

Robert Horne is the somewhat modest and unassuming driving force behind The Golden Apple Trust. His vision and dedication in keeping this important piece of aviation history alive and in the public eye is not to be underestimated. Robert's support is absolute and vital; without him G-SABR simply would not fly.

My fellow pilots: Cliff Spink, Dave Harvey and Keith Dennison who have flown the Sabre brilliantly and professionally. This jet-age icon could not have been entrusted to better hands.

Willem Felger: Golden Apple Operations Ltd's manager, facilitator and the one responsible for many of those unseen activities that are necessary to put the aircraft in the public eye.

Colin Swann, Joe Kennedy, Merill Gilley, George Francis, Andrew Foster (along with others in the hangar, on the airfield and in the office who are too numerous to mention) from the Duxford-based Aircraft Restoration Company. They are dedicated individuals from a superb team who all share the same passion for this aircraft. Over the years the team has overcome many, many complex challenges to ensure the aircraft remains serviceable.

Much help has been forthcoming from the USA as well as from closer to home. Ben Hall and Jim Larsen in Seattle have helped me tell the story of how 178 was brought back to life in the early '70s. Sabre historians Larry Davis (editor of *Sabre Jet Classics*, published by the F-86 Sabre Pilots Association) and especially Duncan Curtis (author of many books and numerous articles about the Sabre) have provided a great deal of material for this book. Both of these historians have forgotten more about the Sabre than I could ever know and have been kind enough to correct many of my historical inaccuracies. Duncan Curtis also kindly and expertly proofread the text, for which I am extremely grateful, as well as providing some excellent graphics and photographs.

Many of the photographs in this book were taken by amateur photographers who all share a passion and enthusiasm for historic aircraft. Special thanks go to the following for allowing their images to be used: Andreas Zeitler, Andrew Critchell, Gary Chambers, Mark Wright, Colin Norwood, Col Pope, Klaas Reinder Sluijs, Paul Brennan, Huw Hopkins, Tim de Groot, Gary Brown, John Myers, Martin Stephen, Gareth Horne, Nick Blacow, Emmanuel Perez, Matt Clackson, Jenny Coffey, Malcolm Clark, Paul Morley, Jim Groom, Alistair Bridges, Chris Muir, Walter van Bel, Paul Sels, Gerry Hill, Neil Cotton and Mike Hall.

Any unaccredited photographs are either of unknown copyright or by the author.

And finally to all our families who support what we do. Without them we could never be properly involved, and consequently this aircraft would never be seen in the air.

Introduction

The North American Aviation (NAA) F-86 Sabre was the world's first operational jet-fighter designed from the outset with swept-wings to achieve high-speed performance. It also featured an axial-flow jet engine, a swept-tail and many other innovations, which included powered flying controls and a highly ergonomic cockpit with outstanding visibility.

Despite design origins emerging during the Second World War, the Sabre led a long and productive life over which it was continuously developed. Originally intended as a high-altitude day-fighter, it was subsequently redesigned into an all-weather interceptor (F-86D) and a fighter-bomber (F-86H).

F-86A, F-86E and F-86F Sabres all took part in the Korean War. The aircraft played a pivotal role in this conflict by winning back air superiority for the United Nations (UN) allies, going head-to-head with the impressive Mikoyan-Gurevich MiG-15 Fagot. The Sabre is ranked, along with its illustrious Second World War ancestor the P-51 Mustang, as one of the great fighter aircraft of all time.

This manual mostly concentrates on the F-86 day-fighters (the 'A', 'E' and 'F'). The subject machine is F-86A USAF serial number 48-178. This aircraft is currently operated on a UK permit-to-fly basis by Golden Apple Operations Ltd and allocated the British registration G-SABR. Based at the Imperial War Museum, Duxford and maintained by the Aircraft Restoration Company (ARC) she participates in 25 to 30 public air shows a year in both the UK and near Europe.

This aircraft is the only example of the 'A' model (the 72nd production aircraft) that remains airworthy and is the world's oldest flying jet aircraft.

BELOW Powerful speed-brakes help to slow down 48-178 as she joins up with the camera ship over Cambridgeshire in 2009. The excellent all-round visibility and pilot's high sitting position are evident in this shot. The camera ship was another Duxford-based North American type – the T-28 Trojan, flown by Ray Corstin. The Trojan was built in 1951, making it a few years younger than the subject! *(Richard Paver)*

Chapter One

The Sabre Story

———●———

The Sabre was conceived in a time of war. It was in these troubled times that the constant pursuit for greater performance and capability led to a period of rapid advancement in weapons, and in aviation especially. Post-war innovations in engineering, materials, aerodynamics and propulsion evolved at such an unprecedented pace that many new aircraft designs were often obsolete while still on the drawing board. The F-86 was an exception to this.

OPPOSITE One of the three XP-86 prototypes on an early test flight over the Mojave Desert. The Sabre was the shape of the jet age. *(North American Aviation)*

In the light of new discoveries, painstaking research and true invention, what had started out as an unremarkable straight-wing jet-powered fighter took on a completely new and revolutionary shape with the performance to match. The Sabre was the torchbearer of the swept-wing concept and it was endowed with a purity of design never to be repeated. It gave to the world the first truly transonic swept-wing fighter and with it the iconic shape of the 'jet age'.

The Sabre remains the most produced Western jet-fighter with nearly 10,000 of the series being built and was the first jet-powered fighter operated by many countries. It remained in active front-line service (with Bolivia) until 1993. The aircraft was produced in twenty different variants (including the Navy FJ series known as the Fury), with five different engines. Only the MiG-15 is believed to have been more widely produced with over 12,000 of these Russian fighters being built. Licensed foreign production of the MiG perhaps raised the total produced to over 18,000.

During its long service life, the F-86 served with the air forces of 34 different countries, including the USA. Here two production lines were established and four foreign countries built the aircraft under licence.

BELOW This diagram, taken from a 1950 NAA publication commissioned by the Air Force to brief the new jet-age pilots, shows the performance improvements brought about by the jet engine and then later by swept-wings.

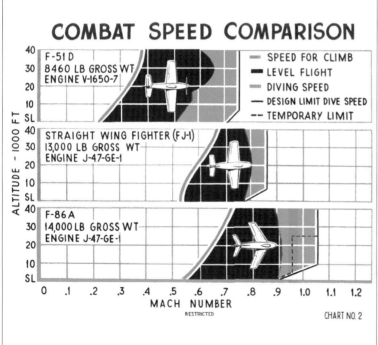

The origin of jet power

By the end of the Second World War piston-engine technology had reached its zenith and opportunities to further improve performance were somewhat restricted by the laws of aerodynamics governing how propellers behave in high-speed flight. Other methods would need to be found to overcome this natural barrier to higher speed.

Various alternatives were available and had been experimented with. One such technology was the rocket engine, which overcame the speed limitations of the propeller, but at the cost of very limited range together with difficult and often dangerous operational problems, particularly with regard to the handling of the volatile rocket fuel and combustion chemicals. Then there was gas-turbine technology, a method of propulsion that had the potential to take aircraft performance to new heights (although not all the 'powers-that-be' realised it at the time). The principle was simple. Air is drawn into the engine and compressed, fuel is added and ignited and the subsequent rapid expansion of hot gases provides the driving force, or put another way, Newton's third law of motion. The expanding gases could also be used to drive the compressor via a turbine thus creating a reciprocating engine capable of sustained performance. The gas turbine also promised greater power, efficiency and lightness. Apart from the spinning turbine/compressor there were virtually no moving parts, so unlike complex piston engines that had literally dozens of valves, pistons, camshafts, crankshafts and other geared components thrashing about under great stress, the jet engine would be more reliable too.

Performance of the earliest jet-powered machines was not much better than contemporary piston-engined aircraft, but they were important first steps that would ultimately lead to an entirely new concept in propulsion and one that would revolutionise aircraft design.

With the end of hostilities in 1945 the stage was set for a new chapter in aeronautical advancement that would exploit the capabilities of the jet engine. Aircraft such as the Messerschmitt Me262 clearly demonstrated the possibilities and were quickly swept up by the Americans,

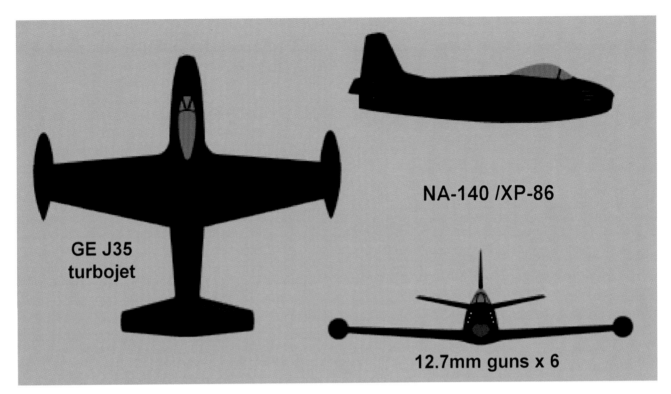

GE J35
turbojet

NA-140 /XP-86

12.7mm guns x 6

the British and the Soviets. Many Me262s were found in readily repairable condition and were confiscated. These aircraft were extensively studied and greatly helped with the development of early Western and Soviet jet-fighters.

Straight-wing XP-86

Before the end of the Second World War NAA, which had made a name for itself with the success of the T-6 Harvard, the P-51 Mustang and the B-25 Mitchell, had begun to look to the post-war future of flight, namely jet-powered aircraft. This work had led to a design for the US Navy of a straight-winged jet-fighter and thanks in part to the advanced nature of this Navy project, NAA were put under contract by the USAAF (United States Army Air Force) in 1945 to produce a new similar jet-fighter. This was the project that would eventually lead to the F-86 Sabre.

Previously NAA had experimented in a variety of ways to increase the performance of the Mustang, including lighter weight, rocket assistance and jets. A rocket was experimentally installed in the after-cooler area of a P-51D and this allowed for a speed of 515mph, but only for about a minute. Thoughts were then given to a very radical P-51 with a propeller, a separate jet

engine and wings that were 'swept' forward. It was found that 'aft sweep' gave wing-tip stall problems, so forward sweep was studied first. Luckily for all involved, the forward swept-wing P-51 never went beyond the drawing board as it was concluded that the wing would suffer from a twisting divergence under load. This highly dangerous and unstable handling trait could not be countered using the then current construction methods and materials. (Much later in 1984 the Grumman Aircraft Corporation would build two X-29 Forward Swept Wing Technology Demonstrators to explore the concept, although this was only possible by the use of advanced composite construction and by employing a computerised fly-by-wire flight control system that overcame the aircraft's inherent instability problems.)

NAA's engineers eventually acknowledged that the future direction lay in a jet-powered aircraft. The required attributes of an aerofoil selection for a new jet-propelled high-speed fighter design were considered and it was concluded that a very thin wing would be essential to avoid compressibility shock. It was important to delay the onset of this compressibility (or wave drag) because it could instantly reduce the lift produced by the wing, bringing no end of handling and stability problems. However, as well as lift and

ABOVE XP-86 prototype of 1944 showing the initial straight-wing configuration. *(North American Aviation)*

Allison J35 gas-turbine engine. The project, by then known as the XFJ-1 (NA-134), was a design intended for the US Navy, which ordered three prototypes on 1 January 1945.

The USAAF also became interested in the Navy project and NAA were awarded a contract to build three further prototype aircraft with the USAAF designation XP-86 (NA-140). The XP-86 did not require the structural strength of a carrier-based aircraft and so compared to the XFJ-1 it was much slimmer and lighter. It differed considerably from its Navy cousin with a refined fuselage shape and deletion of many of the design features intended to aid low-speed performance as would be required for aircraft-carrier operations. The XP-86 wing had the same planform as the XFJ-1, but the aerofoil section was much thinner. On both the upper and lower wing surfaces were dive-brakes, the concept borrowed from the A-36A version of the Mustang. The fuselage had a much higher fineness ratio than the XFJ-1, and the intake was oval in shape. Power was the same for both aircraft – the GE J35 with an eleven-stage axial flow compressor giving 4,000lb of thrust.

The straight-wing XP-86 would have had a pressurised cockpit (the XFJ-1 did not), hydraulic elevator and aileron boost, and wing-tip fuel tanks that could be jettisoned in emergencies. Armament was to be six 0.50-calibre M3 machine guns in the nose (the standard for Army Air Force aircraft at that time). Although Soviet (and German) designers had long focused on cannon armament as the most effective way to bring down enemy bombers, US fighter armament philosophy in the late '40s and early '50s still held to the principle, as in the Second World War, that using heavy-calibre machine guns was more effective due to their vastly superior rate of fire compared to cannon.

Under the wings, a pair of pylons could hold up to 2,000lb of bombs, drop tanks or eight 5in HVAR rockets.

On 20 June 1945, the mock-up of the XP-86 was unveiled at NAA's Inglewood plant. Presented in gloss pearl grey paint, the design

drag characteristics, the requirements for a thin wing raised concerns about structural integrity. The conflicting problems of shape and structure were compounded by a lack of aerodynamic data for a wing section with such small thickness/chord ratio.

On 22 November 1944, NAA initiated a design study for a jet-fighter proposal. The design was straightforward in all respects, and indeed used many P-51 technologies, but it was powered solely by a gas-turbine engine. The flying surfaces, the wings and the tail were very similar to those found on the latest P-51 designs and the wing was of the latest laminar flow design, with straight leading and trailing edges. The fuselage was short, rotund and very smooth. The nose was open to provide an intake to feed air to the

was quickly approved by Army Air Force officials. Photographs of the mock-up show that the aft section of the fuselage had its engine break point about midway through the wing root chord. The fuselage was much sleeker than the somewhat stocky XFJ-1.

The XP-86 was by now a contender for the USAAF General Operational Requirement (GOR) that called for a day-fighter of medium range that could operate in both the escort and fighter-bomber missions, with a top speed in excess of 600mph. This item in the GOR was considered by many to be outside the scope of any jet designs at the time and clearly something would need to change in the basic XP-86 design in order to satisfy it. In the original straight-wing configuration the XP-86 was unremarkable and in some regimes it could not outperform the P-51 Mustang and offered very little advantage over the Lockheed P-80 Shooting Star then entering service. NAA officials knew that the lack of speed would eventually kill off the XP-86 programme unless an engine with more power was developed or some other way to reduce the drag encountered at speeds over 500mph could be found. The advantages (and stability shortcomings) of sweeping the leading edge of the wing to reduce drag rise were well known and so it was that an aerodynamic solution to address the performance shortfall became the focus of the design team.

Contained within the captured German research material was considerable data concerning the use of wing leading edge movable surfaces, i.e. 'slats', that might provide a solution to the instability problems. Based on their own research and by studying these German papers, the design group at NAA was able to convince its superiors that the stability problems associated with wing sweep could be solved and therefore a swept-wing XP-86 would be able to achieve the speed requirement of the GOR. Therefore, on 14 August 1945, NAA received a research and development grant to develop a swept-wing version of the XP-86 and to examine a slat concept as a possible solution to expected stability problems.

A scale model of a swept-wing XP-86 was built and subsequently wind-tunnel tested. The results were exactly what NAA and the USAAF

GOR Requirements, May 1945

1　Day-fighter of medium range.
2　600mph top speed.
3　Dual role as fighter-escort and fighter-bomber (dive-bomber).
4　Pressurised cockpit with ejection seat and 360-degree all-around vision for the pilot.
5　Thin laminar flow wing.
6　Hydraulically controlled (power boosted) elevators and ailerons.
7　External fuel storage that can be jettisoned for combat.
8　Computerised (radar-ranging) gun sight.
9　Six 0.50-calibre M3 nose-mounted machine guns w/267 rounds per gun.
10　Under-wing shackles to carry up to 2,000lb of bombs, external drop tanks or eight 5in HVAR rockets.

had been looking for. The swept-wing lowered the drag rise and compressibility enough that it easily brought the XP-86 into the 600mph+ requirement, even using the then available gas-turbine technology. Moreover, the leading edge slats appeared to solve the low-speed pitch instability. Slats were not the only solution to the stability issues, but in 1945 they fixed a problem that was proving difficult to resolve and allowed the design to progress. On 1 November 1945, General Bill Craigie, head of Research & Development at Wright Field, gave NAA the go-ahead for the building of the swept-wing XP-86 and by February 1946 the mock-up received USAAF approval. This was followed on 20 June 1946 by a formal contract for full-scale development. In August engineering drawings were released to the manufacturing division of NAA and the first metal was cut on the XP-86 prototypes. The USAAF was so keen to press on with the new swept-wing XP-86 with its better performance that a contract was awarded on 20 December 1946 for the construction of 33 production P-86As. No YP-86 service test aircraft were ordered as these P-86A-1s would effectively serve as development aircraft.

In addition to the almost full-span leading edge slats, the final configuration of the wing comprised single-slotted high-lift flaps and outboard ailerons incorporated in the trailing edge portions of the wing. The prototype and eventual production models of the Sabre were designed with a 35-degree wing sweep and an aspect ratio of 5.0. After initially being designed as unswept (on the prototype) the sweep-back was also applied to the tail surfaces. Additionally, the horizontal stabiliser was fully trimable to achieve a better balance between low-speed control and high-speed requirements. This planform was highly successful and remained mostly unchanged for all Sabres until the introduction of the 6-3 hard wing on the F-86F seven years later.

The decision to radically redesign the XP-86 was a brave one for the project team. It is true that the straight-wing XP-86 project was doomed without the new wing, yet because this was such new technology the risks were considerable. Costs incurred on the straight-wing jet-fighter were absorbed by the company, although some of this would be recovered through the Navy decision to go ahead with the straight-wing design that would become the FJ-1 Fury. Incorporation of the swept-wing would put the XP-86 programme back by six months compared to the Navy's XFJ-1, but the Air Force was willing to make this sacrifice knowing that the massive performance gains justified the wait. In the event, the XFJ-1 took to the air for the first time on 27 November 1946 and the first XP-86 flew nearly a full year later.

It wasn't only NAA that pressed ahead with swept-wing technology. The entire US industry embraced the notion including George Schairer, of the Boeing Company, who after the Second World War went to Germany with von Karman and Robert Jones (an early National Advisory Committee for Aeronautics (NACA) proponent of swept-wings) to investigate German data on the technology. Schairer was very enthusiastic about wing sweep, proposing that Boeing incorporate it into the new XB-47, and noting that this information should be made available more widely to the US aeronautical industry.

Interestingly, the Soviet designers did not use leading edge slats to solve the same stability problems that they had encountered during the development of the MiG-15. Instead they used a combination of wing fences and structural washout to achieve acceptable handling. Washout is a feature of wing design intended deliberately to reduce the lift distribution across the span of the wing. The MiG-15's wing was designed so that angle of incidence (angle to the fuselage) was higher at the wing roots and

then decreased across the span, becoming lowest at the wing-tip. This was done to ensure that, at the stall, the wing root stalls before the wing-tips, providing the aircraft with continued aileron control and some resistance to spinning. The MiG-15's wing fences were principally added to reduce the span-wise flow of air and thereby alter the lift distribution to reduce lift-induced drag. These solutions, while acceptable, were not as refined as those on the Sabre and consequently contributed to the MiG-15's inferior slow-speed handling.

XP-86 swept-wing development

As NAA's engineers brought the XP-86 wing design to its final shape the slat design remained a problem until an entire Me262 wing was flown in from Wright Field. These mechanisms were disassembled and although NAA designers were not overly impressed with the design, considering it crude, a modified slat-track mechanism was fitted to the XP-86 wing, using the Me262 slat lock and control switch. As on the Me262, deployment of the slats was automatically initiated by aerodynamic loads at the correct angle of attack acting at the leading edge of the wings. Although not perfect, it was a solution that worked. NAA expended a lot of effort in improving on the German design and eventually an entirely new and efficient slat design with a superior slat-track mechanism was developed. However, the first seven production P-86s did feature a slat lock and control switch that was a direct copy of the one fitted to the Me262 wing.

Further investigation also led to the original wing-mounted dive-brakes being removed. Tested on the XFJ-1, the design was in fact too good for its intended use and there were concerns about the aircraft being uncontrollable should only one side deploy. Shortly before his death, M.M. 'Mac' Blair, a key aerodynamicist with NAA during the XFJ-1/XP-86 years stated, 'It [the XFJ-1] had the most phenomenal speed-brakes in the world. These were the 6-foot long 4x4s that rotated out from each wing panel (one above and one below). After much testing, two insurmountable problems led to [their] demise. First the pilot's eyeballs

LEFT First Prototype XP-86 on rollout at Inglewood. *(North American Aviation)*

BELOW Taken at one of Duxford's air shows in 2008, the Fighter Collection's TF-51 Mustang keeps up with Golden Apple's F-86A, but only just. It's incredible to realise that the first flight dates of these two aircraft in prototype form (XP-51B, 30 November 1942 and XP-86, 1 October 1947) are separated by less than five years. Both aircraft are based at the Imperial War Museum, Duxford, UK. *(Dr A. Zeitler, www.flying-wings.com)*

separated from their sockets due to the instantaneous negative 1 'g' deceleration, and second, above about Mach 0.6 the buffet was so intense that the airplane was nearly uncontrollable.'

Instead, NAA settled on a speed-brake design that comprised two rearwards-opening concertina-type panels on the rear fuselage and a third similar panel under the mid-fuselage just aft of the wing.

As the XP-86s were being built, wind-tunnel testing continued and revealed that the concertina-type speed-brakes would be unworkable in service, as Mac Blair recalled, 'It became obvious that the original [XP]-86 brake design was unsatisfactory: there was far

Photographed shortly after its arrival at Muroc Dry Lake test facility in September 1947, the first XP-86 reveals its smooth and sleek lines.

too much drag area, the pitch trim changes would be terrible, and the [rear-hinged] brakes would fail to open with the loss of hydraulics.

'A series of tests with various configurations of speed-brakes on the [XP]-86 wind tunnel model (the FJ never had a wind tunnel model) and flight tests of the XFJ-1 led to a size and location (and understanding) that gave satisfactory deceleration with acceptable trim changes in pitch, and a minimum of buffet.'

This was the version that went into production, comprising a front-hinged panel on each side of the rear fuselage. Hydraulically actuated and rotating downwards upon opening, this design avoided many of the pitch changes associated with the earlier design. Unfortunately, the Air Force did not approve this version until September 1947 and as a result the No. 1 XP-86 still had the concertina-type speed-brakes fitted for its initial flight testing.

The cockpit sat high on the forward fuselage

with a Plexiglas canopy offering excellent all-round vision for the pilot, something not matched in a fighter aircraft until the McDonnell-Douglas F-15 Eagle was unveiled some 25 years later. All radio and radar antennas were enclosed in fibreglass fairings within the overall lines of the aircraft. The engine powering the prototype was the Chevrolet-built GE J35-C-3 rated at 4,000lb static thrust.

On 8 August 1947 the doors of NAA's Inglewood factory opened to reveal the first prototype XP-86, serial 45-59597. The new NAA fighter was given the 'PU' buzz number prefix (which would also apply to production aircraft) and so the prototype became 'PU-597'. During the next month, taxi and brake tests were conducted at Mines Field (adjoining the factory) before the aircraft was disassembled and trucked to Muroc Dry Lake Army Air Base, now known as Edwards Air Force Base, where it was reassembled and all systems retested.

RIGHT **First prototype XP-86 shows its clean lines on the ground at Muroc after delivery by road from the NAA manufacturing plant at Inglewood.** *(North American Aviation)*

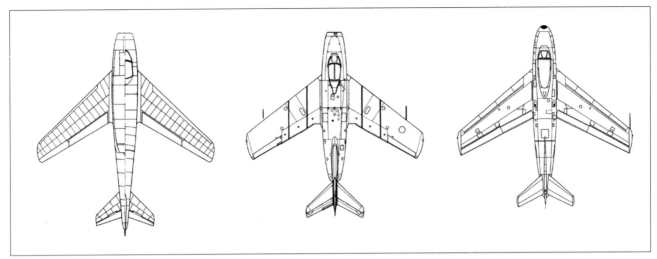

George Welch (Engineering Test Pilot for NAA) took PU-597 on its first taxi run at Muroc on 29 September 1947, with a second run completed the same day. Welch reported that nose-wheel steering was unsatisfactory at speeds above 18 knots and further that the rudder only became effective for steering above 48 knots. Clearly there would be a speed band on take-off where directional control was less than ideal. Nevertheless, the team pressed ahead for a first flight two days later.

The Nazi connection

Throughout the Second World War German scientists had been researching projects that encompassed advanced propulsion technologies (notably rocket and turbo-jet power plants) and aerodynamic developments (swept-wing and tailless configurations) that were all aimed at improving the military potency of their aircraft. In the last months of the war extra resources were allocated to these projects in a desperate effort to regain air superiority for the Luftwaffe. In the closing days of the conflict allied forces advancing from the East and West stumbled upon the laboratories and factories that had been at the centre of these advanced projects, and more significantly captured the scientists together with large amounts of wind-tunnel data.

Consequently in 1945, American, British and Russian aircraft engineers gained access to advanced studies into swept-wing design including data from Volkenröde's high-speed wind tunnels. By studying this data, together with transcripts of the interrogations of captured scientists, US aviation researchers were able to confirm their own discoveries and gain insight into possible solutions to some of the high-speed design difficulties that were proving problematic. In particular, NAA aerodynamicists were able to determine that satisfactory control and stalling characteristics could be achieved by using slats, and that control at high subsonic Mach numbers could be made acceptable by thickening the trailing edge of the wing.

The advantages of sweeping wings had first been enunciated by a young German aerodynamicist, Adolf Busemann, at an Italian-hosted conference in 1935. Ignored at the time by all but Germany, the swept-wing was independently rediscovered by American scientist Robert Jones, working at the NACA's Langley laboratory in Virginia in 1944. Jones had already discovered the benefits of sweeping wings to delay transonic drag rise and shock-wave formation, but these advantages were not without their complications and the data (supplied in 1945 by the Central Intelligence Agency at Langley) offered some answers to the stability problems that were being encountered at high lift values, particularly at slow speed.

The Sabre was the world's first aircraft designed with a swept-wing deliberately to attain a high-speed advantage. Germany had many swept-wing projects on their drawing boards, but it did not field a true swept-wing aircraft. The Messerschmitt Me262 was, like the Douglas DC-3, basically a straight-wing aircraft with pronounced leading edge taper, and the rocket-powered Me163, which used a swept-

ABOVE The final planforms selected by NAA for the Sabre (right) and Mikoyan-Gurevich for the MiG-15 (middle) look remarkably similar to the Nazi experimental Messerschmitt Me P.1101 (left) (dated 8 November 1944). Comparison of the outlines illustrates that aerodynamicists from the USA, the Soviet Union and Germany were all independently coming to the same conclusions about optimum swept angle. Swept-wings were not a unique innovation, but the research carried out by German scientists helped to solve some of the instability problems inherent in the design.

wing, had such abysmal flying qualities that it killed more pilots in 'flying' accidents than as a result of enemy action.

There can be no doubt that the problems encountered would have eventually been solved, but it is also true that the allied victors were able to advance their own understanding of the technologies required to develop high-performance aircraft by studying the work done by German scientists, engineers and aerodynamicists.

XP-86 maiden flight

On the morning of 1 October 1947 George Welch taxied the XP-86 to the end of the runway at Muroc, released the brakes and pushed the throttle forward; 3,000ft down the runway PU-597 lifted smoothly off the dry lake bed for the first time to the applause of the personnel observing.

It has been (wrongly) suggested that George Welch went supersonic on the maiden flight of the XP-86, thereby beating Air Force Captain (Capt) Charles 'Chuck' Yeager, the official holder of this record in the Bell X-1, by two weeks. The XP-86 was certainly capable of supersonic flight and frequently achieved this status during the test programme, albeit in a dive, but the evidence to suggest that this event occurred on the very first flight is non-existent and indeed there is evidence to prove that it could not have happened.

The first flight of PU-597 was dogged by landing gear problems. It is known from

the flight test report and verified by the accompanying chase plane that after climbing to altitude both main gears failed to retract completely, while the nose gear showed an unusually rapid retraction. Several cycles of the landing gear were made in an attempt fully to raise the main gear, but without success. It was then noted that the nose gear failed to extend beyond approximately 20 degrees, although hydraulic accumulator pressure was normal. In an effort fully to extend the nose gear the emergency down system had also been employed, but without success.

George Welch then flew overhead the test crew who also confirmed that the nose gear appeared to be extended by only about 45 degrees. With fuel by now getting very low Welch declared that he would stay with the aeroplane and attempt to land on the dry lake bed, not one of the runways. He made a smooth, very nose-high approach. Touching down, George let the airplane coast along with no brake application. As the aircraft slowed and the nose started to lower the nose gear swung forward and locked into place and a disaster was averted. It was later discovered that the hydraulic ram on the nose-gear retraction system was too small. The nose leg folded to the rear and thus extended into the airflow and the ram was simply not strong enough to overcome the air loads. The immediate fix was to use two of the original cylinder/pistons, then replace them later with the correct specification.

Instead of the planned-for 10-minute test-hop, Welch had been airborne for 1 hour and 18 minutes. Over the years a number of sources have stated that Welch had experienced nose-gear problems at the end of the first flight and that he therefore had had the opportunity to break the sound barrier 'before' he experienced any problems. Knowing that Welch had encountered problems as soon as he raised the XP-86's undercarriage for the first time, it is obvious that he would have been occupied thereafter with more pressing issues than trying to break the sound barrier. He actually spent almost all of the first flight trying desperately to coax first the main landing gear and then the nose leg into a safely locked-down position. Because of these problems PU-597 then had its landing gear fixed in the

LEFT The second prototype XP-86 wearing the designation 'PU-598'; this was later changed to 'FU-598' after 1948 when the USAF naming convention changed from 'P' for 'Pursuit' to 'F' for 'Fighter'. *(North American Aviation)*

down position until several weeks after Chuck Yeager's historic flight in the X-1, and so there was no other opportunity for Welch to beat him to the sound barrier.

Conspiracy theorists will always contend that NAA had been forced to cover up certain details of the first flight. It is true that when the XP-86 did finally go supersonic the company was indeed instructed to keep this information from getting out, although the event was properly recorded in the relevant NAA Flight Test Reports. About a year into the test programme supersonic flight in the XP-86 would become common knowledge, by which time it was no longer seen as such a significant achievement.

To argue or endlessly speculate about who broke the sound barrier first is to overlook a very important and crucial point. Supersonic flight from the Sabre onward would be routine. Here was an aircraft that could be taxied out to the end of the runway, be taken airborne, climb to altitude, dive to a speed greater than Mach 1.0,

LEFT First prototype XP-86 No. PU-597 in flight. Note the leading edge slats (made up of three sections per side), which are partially and unevenly extended. Later, and for production Sabres, these slats would be bolted together so as to move as one piece. *(North American Aviation)*

land, return to the pan and then do it all again just as soon as it could be refuelled. The X-1 was never able to do this. The X-1's place in history is safe and justified, although the Sabre shows us that the rocket plane would never be a practical or operational solution. The Sabre story also demonstrates just how quickly things were advancing in the world of high-speed flight in the late '40s. The XP-86 may not have been the first, but it was the future, and thereby revealed the X-1 to be a pure research aircraft that would never have any practical application in the real world. It could also be argued that the XP-86 actually contributed more to our knowledge of high-speed flight than the X-1 and was thus a better experimental platform.

Aside from the early landing gear problems, Welch's feedback on the flying characteristics of the new machine was generally favourable and his only concern was that the prototype aircraft did not have enough power. With only 4,000lb of thrust, the XP-86 had a rate of climb of just 4,000ft/min. Production P-86As would have 5,000lb of thrust from the J47 engine so no one was too concerned. Misgivings about high aileron break-out forces (the amount of effort required to move the control column from a static condition) were also aired by Welch, and eventually a modified aileron bell-crank torque-tube with needle-roller bearings sorted this out.

Phase II flight tests flown by Air Force pilots

ABOVE Prototype XP-86 in which George Welch conducted the first flight on 1 October 1947. This was a revolutionary new shape in the sky and even with the limited thrust available from a non-production standard engine offered much better performance than anything flying at that time (X-1 rocket-ships aside). *(North American Aviation)*

BELOW Pre-1948 picture of XP-86 prototype No. PU-597 on an early test flight. The aircraft still has the concertina-type speed-brakes and small test probes under the cockpit. *(North American Aviation)*

ABOVE First prototype XP-86 painted in overall gloss grey and now wearing the 'FU' buzz numbers. The repaint was completed after the aircraft had been modified with production-type dive-brakes and rear fuselage. It was thought that the gloss paint would make the aircraft faster; however, the resulting additional weight of the paint actually made the aircraft slower. *(Peter Bowers)*

ABOVE First Prototype XP-86 in flight. Note special test pitot probes fitted to each wing-tip and the top of the fin. The prototype canopy is also clearly seen – the profile was changed for production aircraft and included a navigation light at the rear. *(North American Aviation)*

BELOW All three prototype XP-86s in line-abreast formation high above the Mojave Desert. By the time this photograph was taken all three of the aircraft had been handed over to the USAF. All three aircraft are also now fitted with production-standard rear fuselages. *(North American Aviation)*

began in December 1947. In early 1948, XP-86 prototypes No. 2 (45-59598) and No. 3 (45-59599) were finished and joined the test programme. They were different from the No. 1 aeroplane, as well as from each other. Both No. 1 and No. 2 had different fuel gauges, a stall warning system built into the control stick, a bypass for emergency operation of the hydraulic boost system and hydraulically actuated slat locks. XP-86 No. 1 was the only one with an on-board fire extinguisher. The No. 3 was the only one with automatic slats that opened at 135mph, and full armament.

F-86A Enters Service

On 28 December 1947 a subsequent order for 190 P-86Bs followed the initial contract for 33 P-86As. The P-86B was never built (see explanation elsewhere) and this order was changed to a further 188 P-86As. The production aircraft were to be powered by the General Electric J47-GE-1 jet engine developing 5,200lb of thrust.

On 16 September 1947 Congress made the Air Force a separate branch of service from the Army. Consequently, in June 1948 the new US Air Force re-designated all Pursuit aircraft to Fighter aircraft, changing the service prefix from P to F. Thus all XP and P-86A aircraft were renamed XF and F-86A respectively. Deliveries to the USAF in three initial batches began in February 1949 and the name 'Sabre' was officially bestowed on the aircraft after a USAF-sponsored naming contest. All F-86As were to be manufactured at Inglewood, California (now the site of Los Angeles International Airport (LAX)). The aircraft carried the company suffix 'NA', as in F-86A-1-NA. Production began on the assembly line as the last FJ-1 Fury was completed for the US Navy.

Compared to the XP-86 prototype the F-86A's wingspan was increased by 1in to 37ft 1in and the empty weight rose to 10,093lb, yet thanks mostly to the more powerful engine

BELOW First prototype XP-86 banks away from the camera ship. The concertina-type speed-brake on the underneath of the aircraft is inoperative and has been 'speed-taped' to the seal the gaps. *(North American Aviation)*

LEFT The 393rd F-86A (the serial number 91178 shows this to be the 1,178th USAF aircraft from FY 1949 budget) in production at NAA's Inglewood Plant, now the site of LAX (Los Angeles International Airport). *(North American Aviation)*

CENTRE The artist's rendering of the XP-86 swept-wing design shows some features that were deleted or changed on the production aircraft. The dive-brakes on the rear fuselage are hinged at the rear with the actuating arms behind the dive-brake panel; on production aircraft the brakes were hinged at the front. Also seen here is the under fuselage dive-brake which was deleted entirely on production aircraft. *(North American Aviation)*

BELOW LEFT F-86A Sabres in production at NAA's Inglewood Plant. Such was the demand for the Sabre that with the production of the F-86F and naval FJ-2/3 Fury a second production line was set up in Columbus, Ohio. *(North American Aviation)*

BELOW The first production F-86A-1 in flight. Most A-1s were essentially trials and test aircraft. At this time the aircraft has yet to have any guns or the gun panels fitted. Note the curved windscreen of the A-1; A-5s, A-7s and early Es were fitted with the 'vee' windscreen. *(US Air Force)*

the maximum speed at sea level greatly increased to 679mph, an improvement of 80mph. The Sabre could climb to 40,000ft in 10.4 minutes and then cruise at 533mph. The service ceiling stood at 48,000ft with a clean range (internal fuel only) of 660 miles. For an aircraft produced in the late '40s these performance figures (apart from the range) were truly outstanding and represented an enormous increase in capability over early jet aircraft and the last generation of piston-engine fighters.

F-86As were built in production blocks from 1947 to 1949. They used USAF serial numbers 47-605 to 47-637, 48-139 to 48-316 and 49-1007 to 49-1339: a total of 554 'A's were eventually built. Engine production delays held back the last F-86As from delivery until March 1949.

The Sabre refined

The F-86A was not perfect. Given that swept-wings were a totally new concept it is not surprising that the learning curve for NAA was steep. In some respects the F-86A was an experimental type with some inherent faults and it is perhaps no surprise that the F-86E that followed incorporated many refinements.

Aerodynamically, and in handling terms, the major characteristic that wasn't quite right was poor pitch control at high transonic and supersonic Mach numbers. This was due, in effect, to a blanking of the elevator surfaces behind the shock wave that formed along the tail-plane leading edge. From the F-86E onward this was solved by incorporating a movable stabiliser with a linked elevator. The final versions of the F-86 were further improved in the area of transonic pitch response by incorporating an all-moving, slab-type horizontal tail with no elevator. This arrangement was so successful that it was to become standard on future transonic/supersonic fighters.

Also in the area of flight control, all Sabres from the F-86E onward had dual system irreversible, fully powered hydraulically actuated controls with artificial feel. The more complex controls of these later variants aided in eliminating such instabilities as aileron and rudder buzz, in addition to permitting maximum deflection of the control surfaces without requiring excess physical effort on the part of the pilot.

Environmental control in the cockpit consisted of air conditioning, heating, windshield demisting and pressurisation. In addition, the pilot sat under a cockpit canopy that could be jettisoned in flight by an explosive charge and was seated on an ejection seat that could be catapulted clear of the aircraft by pyrotechnic charges. The ejection seat was a new concept to most pilots converting to the F-86. Anecdotes suggest that many did not trust the seat explosives and deliberately flew with the safety pins fitted! The ejection seat was deemed essential given the high speeds that might be encountered and the high-placed tail-plane. The seat was very basic by modern standards.

The F-86A was superceded by more capable Sabre models and then these in turn by the 'Century Series' aircraft. After removal from front-line operations many Sabres enjoyed second careers serving with state-based Air National Guard units. For longer still a number remained in use as target drones and others were employed for various flight-test purposes: at least one aircraft manufacturer used an F-86 as a chase plane until very recently. The handful

of Sabres that remain airworthy in 2011 are all privately owned.

Early Sabres that never went into production

The F-86B

The F-86B was intended to meet an Air Force requirement for bigger tyres on the Sabre. The fitment of larger tyres would have meant increasing the width of the fuselage by 7in. However, advances in tyre and brake technology by 1949 allowed the Sabre to retain its original tyre size. Therefore, a planned 190 F-86Bs became 188 more F-86A-5s and 2 F-86Cs. No F-86Bs were built.

The F-86C

The F-86C, or Model NA-157, was NAA's reply to the Air Force's request for a 'penetration fighter interceptor' or long-range escort-fighter. Its design was initiated on 17 December 1947. The F-86C used the same wings and tail surfaces, but otherwise was a radical redesign and much larger than the F-86A. It

had a bullet-shaped nose with the air intakes placed on the side of the fuselage. There were now six 20mm cannons and the addition of radar in the nose. The F-86C was powered by a J48-P-1 engine with afterburner that developed 8,000lb of thrust. This engine was an American-built version of the British Rolls-Royce Tay assembled by Pratt and Whitney. As the F-86C was so radically redesigned to accommodate this engine, it was re-designated YF-93A. The aircraft was bigger and heavier, requiring dual main wheels and a larger internal fuel supply. The F-86C/YF-93A competed with the McDonnell XF-88 and Lockheed XF-90, and 188 of NAA's entry were ordered in June 1948. The first of two YF-93As built was flown by George Welch on 25 January 1950. However, by then the programme had already been cancelled and no further YF-93As were assembled. The Air Force instead ordered more bombers. The NACA purchased both YF-93As for testing lateral air intakes and they were assigned to NACA's Ames Test Centre near San Francisco, California. Neither the F-88 nor the F-90 went into series production either.

ABOVE The 1st Fighter Group at March Air Force Base, California, was the first unit to receive the F-86A in February 1949. The troublesome flush-fitting gun muzzle doors were only fitted to early F-86As. These aircraft are fitted with the large 252-gallon 'ferry tanks', although due to weight restrictions only 206.5 gallons were allowed to be loaded.

Test Pilot George Welch

Geeorge Welch (10 May 1918–12 October 1954) was a remarkable man. Apart from his test-flying exploits, Welch is best known for being one of seventeen USAAF fighter pilots able to get airborne to engage Japanese forces during the attack on Pearl Harbor (Welch was portrayed in the 1970 film *Tora! Tora! Tora!* by Rick Cooper). Welch was nominated for the Medal of Honor for his actions and Air Force Chief, General (Gen) Henry H. 'Hap' Arnold, supported the nomination. However, for reasons unknown, intermediate commanders declared that Welch had taken off without proper authorisation and could therefore not be awarded the nation's highest military award. Instead, he received the Distinguished Service Cross.

In the spring of 1944, Welch was approached by NAA to become a test pilot for the P-51 Mustang. With the recommendation of Gen Arnold, Welch accepted. He went on to fly the prototypes of the FJ Fury, and when the F-86 Sabre was proposed, Welch was chosen as the chief test pilot.

Following his work on the XP-86, Welch continued to work with NAA as chief test pilot. Indeed, during the Korean War he worked in theatre developing the Sabre's '6-3' wing. He was also an instructor and reportedly downed several enemy MiG-15s himself while 'supervising' his students. However, Welch's 'kills' were in disobedience of direct orders for him not to engage and credits for the kills were thus distributed among his students.

After the Korean War, Welch returned to flight-testing, this time in the F-100 Super Sabre. Welch became the first man to break the sound barrier in level flight with this type of aircraft on 25 May 1953. However, stability problems with the aircraft arose, and on Columbus Day, 12 October 1954, Welch's F-100A-1-NA Super Sabre, 52-576 disintegrated during a 7g pull-out at Mach 1.55 and Welch was killed. The instability was due to roll coupling that was solved by the F-100 production version being fitted with a taller fin. He is buried in Arlington National Cemetery.

George Welch (10 May 1918–12 October 1954). *(North American Aviation)*

ABOVE **Test pilot and record breaker Maj Robert L. Johnson. Paint erosion on the nose cone suggests that this aircraft was flown fast and low in an abrasive (sandy?) environment.**

Record breakers

The F-86A was by far the fastest combat aircraft in the USAF inventory at the time of its introduction and it was only natural that both NAA and the USAF would be keen to attempt to break the official world speed record to demonstrate publicly this prowess. The Cleveland National Air Races event on 5 September 1948 was chosen as the venue and Major (Maj) Robert L. Johnson of the Air Materiel Command flew an F-86A-1, No. 47-611 for the endeavour. The rules specified that the record attempt would be flown over a closed course, 3km in length, at a height of no greater than 165ft. The pilot had to cover the course twice in each direction during one continuous flight. On the day poor weather and other aircraft breaking into the closed course prevented Maj Johnson from officially setting a new record because only three of the runs were timed. The average speed was 669.480mph, easily breaking the record; it just wasn't official.

On 15 September 1948 at Muroc Dry Lake in California (now Edwards Air Force Base) Maj Johnson tried again and this time made the required number of timed runs. For this flight he flew another F-86A-1, No. 47-608, to an official record of 670.981mph. The previous record was 650.796mph held by the Douglas D-558-1 Skystreak, a Navy experimental aircraft, set on 25 August 1947.

Another notable Sabre record breaker would be Colonel (Col) Fred Ascani flying F-86E-10, No. 51-2721. The Air Force wanted to show off the newest in the Sabre series and Col Ascani was ordered to take two F-86E-10s to the National Air Races, held on 17 August 1951 in Detroit, Michigan, and attempt to set a new closed course record. The 100km closed course was set up and the timers installed. At the end of the day, Col Ascani had a new record with a speed of 635.686mph. Again the record was set in a production aircraft. And to make the point the Air Force included No. 51-2721 in the next shipment of Sabres that were sent to Korea, where it was assigned to the 51st FIG at Suwon Air Base.

The F-86D all-weather interceptor was the first Sabre to have an afterburner installed (J47-33, which produced 7,650lb of thrust), giving this version of the aircraft much more thrust than any previous variant. In the autumn of 1952 the Air Force put on a public demonstration (i.e. for the benefit of the Soviets) of the much greater speed their new interceptor possessed. On 19 November Capt J. Slade Nash, flying F-86D-20 No. 51-2945, took off from El Centro NAS and headed for the Salton Sea, another dry lake bed, where a timed course had been set up. Capt Nash took the F-86D up to 1,000ft, then nosed down and crossed the entry point of the course at 100ft!

Capt Nash exited the course 9.6 seconds later and made a tight 3G turn and headed back for the second run. After four such passes were made and verified, Capt Nash and the F-86D held a new World Absolute Speed Record with an average speed of 698.505mph, breaking the old record held by Maj Robert Johnson and the F-86A back in 1948.

Capt Nash's record didn't stand long. On 16 July 1953 Lieutenant Colonel (Lt Col) William Barns, Air Materiel Command's pilot representative at NAA, would make a record attempt. Lt Col Barns made the FAI-required four passes in F-86D-35 No. 51-6145, breaking the sonic barrier on each pass. His four passes were timed at 720.574, 710.515, 721.351 and 710.350 giving an average of 715.697mph, over 17mph faster than Capt Nash's record flights. Not only was Lt Col Barns' F-86D a production aeroplane, he flew the record course with full combat loading including 24 2.75in rockets in the tray.

Over the lifespan of the aircraft, Sabres held a number of other records. In January 1954 Col Willard Millikan, Commander of the DC Air National Guard, set a west coast to east coast speed mark in a standard F-86F. Yet another coast-to-coast record was set by a California Air National Guard F-86A named 'California Boomerang'.

Records are made to be broken and within a few years all Sabre records had fallen and it wasn't long before the World Absolute Speed Record was well over 1,000mph. However, for at least eight years the F-86 ruled supreme as the world's fastest production aeroplane.

BELOW F86A-1, No. 47-611 flown by Maj Robert L. Johnson on 5 September 1948 to establish a new 'unofficial' world speed record of just over 669mph.

Sabre astronauts

There were a number of National Aeronautics and Space Administration (NASA) astronauts from the Mercury, Gemini and Apollo eras, who together with other key personalities involved in these space programmes, flew F-86s, many operationally during the Korean War.

Colonel (Col) Edwin 'Buzz' Aldrin (Gemini 12, Apollo 11) flew 66 combat missions in Korea; Lieutenant Colonel (Lt Col) Virgil Ivan 'Gus' Grissom (Mercury-Redstone 4, Gemini 3, Apollo 1), who tragically died in a fire during testing for the Apollo 1 mission, flew 100 missions; and Second Lieutenant (2 Lt) Gene Kranz, NASA flight director for Gemini and Apollo and assistant flight director on Project Mercury, also patrolled the Korean Demilitarised Zone.

Senator John Glenn (Mercury-Atlas 6, STS-95), the first American (and third person) to orbit the Earth, was a Marine Corps fighter pilot before joining NASA's Mercury programme. During his second combat tour in Korea on an inter-service exchange programme with the USAF he logged 27 missions in the F-86F Sabre, and shot down 3 MiG-15s near the Yalu River in the final days before the ceasefire. Glenn's aircraft was painted with the title 'MiG Mad Marine'.

Although not seeing service in Korea, astronauts Charles Duke (Apollo 16), Thomas Stafford (Gemini 6A, Gemini 9A, Apollo 10, Apollo-Soyuz), Gordon Cooper (Mercury 9 (Faith 7), Gemini 5), Jack Swigert (Apollo 13) and Michael Collins (Gemini 10, Apollo 11) all flew Sabres, while Richard F. Gordon, Jr. (Gemini 11, Apollo 12) flew the FJ Fury with the US Navy.

ABOVE USAF NAA F-86F-30-NA (s/n 52-4584) in Korea in 1953. This was the aircraft flown by US Marine Corps Major (Maj) John H. Glenn and dubbed 'MiG Mad Marine'. Glenn flew this aircraft during his time as an exchange pilot with the 25th Fighter Squadron, 51st Fighter Wing, in mid-1953. He actually named his aircraft 'Lyn Annie Dave' for his wife and two kids; however, the crew surprised him with a new paint job, a big red M with letters trailing to read 'MiG Mad Marine'.

RIGHT John H. Glenn Jr stands in the cockpit of the F-86F he flew during the Korean War.

First Woman to Break the Sound Barrier

Jacqueline 'Jackie' Cochran (11 May 1906–9 August 1980) was a pioneer American aviator, considered to be one of the most gifted racing pilots of her generation. She was an important contributor to the formation of the wartime Women's Auxiliary Army Corps (WAAC) before which she served with the Air Transport Auxiliary in Britain during the Second World War. Jackie became the first woman to break the sound barrier, which she accomplished in a Canadair-built F-86E flying alongside Chuck Yeager. This occurred on 18 May 1953.

The First British Pilot to fly Supersonic

Test pilot Roland 'Bee' Beamont was the first British pilot to exceed Mach 1.0 when he took the number two XP-86 (PU-598) supersonic on 21 May 1948. Bee was invited to have a go at 'breaking the sound barrier', an event that had proved tragically elusive for British pilots. To him it was such a non-event that he couldn't believe he'd done it! However, observers on the ground heard the characteristic 'ba-boom' and cockpit instruments confirmed the supersonic reality. Back home any publicity concerning Beamont's achievement was 'withheld' for reasons of 'national security' and Bee himself was sworn to secrecy. The British boffins didn't want to validate the data either, perhaps preferring that honour to be postponed until it could be won in a British aircraft.

The British aircraft, British pilot accolade was subsequently to go to John Derry in the experimental de Havilland 108 Swallow, although in truth the achievement was marginal and nearly cost Derry his life. The feat was only possible in a near-suicidal vertical dive with the Swallow virtually uncontrollable and at the limit of its structural integrity. It was a miracle that Derry wasn't killed.

Beamont wrote a highly detailed account of the handling, performance and systems of the XP-86, from which the following conclusions were drawn:

'The [X]P-86 is an outstanding aircraft in that:
1 It can achieve Mach number exceeding unity.
2 It is fully operationally controllable at up to M=0.97 at least above 20,000ft, and gives the impression of being even aerobatically controllable at M=0.9.
3 It is probably capable of development to IAS of 700mph or more.
4 Despite its extremely high performance capabilities, it is an unusually pleasant and straightforward aircraft to fly – possibly more so than most jet-fighters on both sides of the Atlantic.
5 It is not merely a research aircraft, in which field and that of record-breaking it is likely to shine, but is a standard fighter aircraft in production for the USAF who should experience no functional difficulty in operating it.

'I could find no-one who was prepared to suggest the effect of the firing of the six 0.50in machine guns in the nose at M=1.0.'

ABOVE **Chuck Yeager stands with Jacqueline Cochran in the cockpit of the Canadair-built F-86E in which she became the first woman to break the sound barrier.**

LEFT **British test pilot Roland 'Bee' Beamont went supersonic in the XP-86 in May 1948.**

Film star and cultural icon of '50s America

BELOW One of the most famous concept cars of 1951 was the Buick LeSabre. It was designed by General Motors' chief stylist, Harley J. Earl, who was clearly influenced by many of the Sabre's visual attributes. Swept-wings, tail-fins, jet-exhaust nozzles, Plexiglas canopies and engine air intakes all found their way on to production and concept cars penned by him during the period.

As well as introducing the world to new technologies, capabilities and performance, the style and elegance of the F-86 proved to be an inspiration to many. The 'Sabrejet' became a cultural icon of 1950s America with film makers, auto designers and musicians all drawing inspiration from its sleek lines and futuristic styling.

The Sabre was to appear in many Hollywood movies of the time. It featured in *The Hunters,* which starred Robert Mitchum as Cleve 'Iceman' Saville and Robert Wagner as Ed Pell, with an assortment of F-84Fs pretending to be MiGs. *The Hunters* was filmed at Nellis Air Force Base near Las Vegas in 1957. Loosely based on the novel written by James Salter, who actually did fly with the 4th FIG in Korea, the movie had some incredible aerial scenes.

Hollywood film *Sabre Jet*, directed by Howard Hawks and starring Robert Stack, again saw the Sabre immortalised on film. In that movie, Sabres both played themselves and were the 'MiGs'. Once again the film was shot at Nellis Air Force Base in the early 1950s. It documented the story of both the fighter pilots in Korea and the wives they left behind.

Another classic was *The Day the Earth Stood Still*, where a young lad asks the visitor from outer space, 'How fast does a flying-saucer go?', 'Oh, pretty fast', is the reply. The boy then asks, 'As fast as a Sabre Jet?'

In the film *The McConnell Story*, which starred Alan Ladd as Joe McConnell and June Allyson as his wife, 'Butch' McConnell, the Sabres again came from Nellis, while the 'MiGs' were 366th FBW F-84Fs out of Alexandria Air Force Base. *The McConnell Story* is a dramatisation of the life and career of USAF pilot Joseph C. McConnell (1922–54), who served as a navigator in the Second World War before becoming the top American ace during the Korean War. Ladd, who hated flying, filmed his scenes in mock-ups in front of blue screens. The film also helped establish the 'Missing Man Formation' as part of military aviation culture. This is a salute to a killed pilot in which a formation of aircraft overflies the funeral parade as a tribute, whereupon one aircraft breaks away leaving a gap in the formation.

Most famous of all was *Jet Pilot*, which took almost four years to film. The last day of shooting was in May 1953, but the film was kept out of release by Howard Hughes's tinkering (something he was notorious for) until October 1957. The movie starred John Wayne and Janet Leigh, and the men and aircraft of the 1st Fighter Group at March Air Force Base. *Jet Pilot* was reportedly Howard Hughes's favourite film, one he watched repeatedly in his later years.

The Sabre was such an iconic design that over the years countless children's toys, plastic models and TV shows have featured the aircraft.

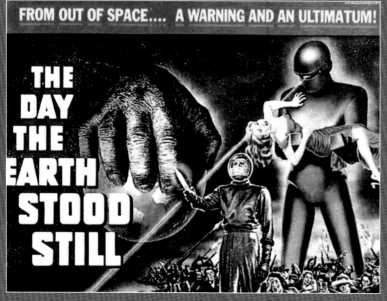

FROM OUT OF SPACE.... A WARNING AND AN ULTIMATUM!

THE DAY THE EARTH STOOD STILL

ABOVE Captain (Capt) Joseph McConnell Jr was the top ace of the Korean War with 16 kills and the subject of the Hollywood movie *The McConnell Story*. He was assigned to the 39th Fighter-Interceptor Squadron of the 51st FIW during the war. After military service McConnell went to work for NAA and was killed testing an F-86H at Edwards Air Force Base, California.

RIGHT FU-897 'The Huff', a Korean War F-86F, was featured as a plastic model by Airfix. This scheme was worn for a time by a Canadair Sabre 5 aircraft flown by Vintage Wings of Canada as C-GSBR.

RIGHT John Wayne starred in the classic Hollywood movie *Jet Pilot*. Although the movie was shot in 1953, it was not released until 1957.

FAR RIGHT Colonel (Col) Walker 'Bud' Mahurin was immortalised as a Sabre-pilot action figure. During the Second World War Mahurin was the first American pilot to become a double ace in the European theatre and was the only USAF pilot to shoot down enemy planes in both the European and Pacific theatres, and the Korean War where 'Bud' was credited with shooting down 3.5 MiG-15s in Korea.

Chapter Two

The Sabre at War

The Sabre story would not be complete without mention of its exploits in battle. At war the Sabre consistently proved itself superior to its opponents even in later life when on paper it had inferior performance. This is surely evidence of how good the initial design of the aircraft was and proof, if needed, that it was a 'pilot's' aeroplane in every sense.

OPPOSITE **F-86A-5s of the 4th FIW at Kimpo in 1951. FU-236 and FU-276 were later shot down by MiGs and FU-158 suffered damage.** *(USAF)*

The Korean War – 1950–53

The Korean War of 1950–53 provided the F-86 with its first exposure to combat. Its main adversary in that conflict was the Soviet MiG-15 Fagot, an interceptor possessed of very similar performance characteristics. Development of the MiG-15 had very much paralleled that of the Sabre and likewise it was one of the first successful swept-wing jets.

The Korean War was a military conflict between the Republic of Korea, supported by the UN, and the Democratic People's Republic of Korea, which was supported by the People's Republic of China (PRC), with military materiel aid from the Soviet Union. The war was the first significant armed conflict of the cold war. The scale and horror of the conflict is often forgotten, there being many different accounts of how many military personnel and civilians died. Estimates varying from between 3 million and 10 million are claimed, the latter being as many as died during the First World War.

The Korean peninsula had been ruled by Japan from 1910 until the end of the Second World War. In 1945, following the surrender of Japan, US administrators divided the peninsula along the 38th Parallel, with US troops occupying the southern part and Soviet troops occupying the northern part.

In 1948 the North established a Communist government and the 38th Parallel became a political border between the two Koreas. Although reunification negotiations continued in the months preceding the war, tension intensified with cross-border skirmishes and raids along the 38th Parallel persisting. The situation escalated into open warfare when on 25 June 1950 North Korean forces invaded South Korea. The Northern forces, equipped with Soviet tanks and trained by the Soviet military, sought to reunify Korea by force. The South Korean military, which was little more than a 'police force' equipped with small arms, was quickly overrun, and fell back in full retreat down the peninsula.

The UN, and particularly the USA, came to the aid of South Korea in repelling the invasion. A rapid UN counter-offensive drove the North Koreans past the 38th Parallel and almost to the Yalu River, causing the PRC to enter the war on the side of the North. The Chinese, aided by the Soviet Union, launched a counter-offensive that repelled the UN forces.

With both North and South Korea sponsored by external powers, the Korean War became a proxy war combining the strategies and tactics of the First World War and the Second World War – it began with a mobile campaign of swift infantry attacks followed by air bombing raids, but had become static trench warfare by July 1951.

The war lasted until an armistice was signed on 27 July 1953. The ceasefire restored the border between the Koreas near to the 38th Parallel and created the Korean Demilitarized Zone (DMZ), a 2½-mile (4km) wide buffer zone

RIGHT Maps of Korea showing the location of the 38th Parallel, the Yalu River and 'MiG Alley'.

between the two countries. Minor outbreaks of fighting across the border and in the disputed Yeonpyeong islands located in the Yellow Sea close to North Korea continue to the present day.

The USA was not well prepared for the conflict. The vast armies fielded in the Second World War had been demobilised and, with the ending of the draft in 1947, the US Army shrank to 550,000 personnel. Much of this force was tied up in occupation duties, with four divisions in Japan, two for a time in Korea and another in Germany. In the summer of 1948 Congress had renewed Selective Service, spurring recruitment and allowing a slight build-up of Army strength to about 600,000.

When the offensive began the North Korean Air Force was equipped with Second World War-vintage Soviet prop-driven fighters, including Ilyushin Il-10s and Yak-9Ps. The vast numerical and technical superiority of the USAF, led by advanced jets such as Lockheed F-80 Shooting Star and Republic F-84 Thunderjet fighters, quickly brought air superiority, thus laying North Korea's cities open to the destructive power of B-29 bombers, which, together with Navy and Marine aircraft, roamed the skies largely unopposed.

The decision to introduce the MiG-15 not only closed the jet-fighter gap, its performance leapfrogged all of the opposing straight-winged jets. The MiG proved very effective in its designed role to attack formations of B-29 heavy bombers, shooting down numerous aircraft.

For many years, the participation of Soviet aircrews in the Korean War was widely suspected by the UN forces, but consistently denied by

the Soviet Union. With the end of the cold war, however, Soviet pilots who participated in the conflict began to reveal the extent of their involvement. It is now known that Soviet aircraft were adorned with North Korean or Chinese markings and their Russian pilots wore either North Korean uniforms or civilian clothes to disguise their origins. For radio communication, they were given cards with common Korean words for various flying terms spelled out phonetically. These deceptions did not long survive the stresses of air-to-air combat, however, and pilots routinely communicated (cursed) in Russian. Soviet pilots were prevented from flying over areas in which they might be captured and which would then reveal that the Soviet Union was officially a combatant in the war.

The Soviets were already training Red Chinese MiG-15 pilots when Communist China

RIGHT A famous photograph of the Torii gate leading to the Sabre flight line at Kimpo Air Base. Although Torii gates had their origins as entrances to Japanese Shinto shrines, this image has become an enduring symbol of the Air Force during the Korean War. *(USAF)*

BELOW F-86As lined up for take-off on a 'mass effort' sortie over Korea. The pilots at the back of this formation must have hoped for a stiff cross-wind to help clear away all that smoke. *(USAF)*

entered the war in support of North Korea. By October 1950, the Soviet Union had agreed to provide air regiments of MiG-15s, along with the trained crews to fly them. Simultaneously, the Kremlin agreed to supply the Red Chinese and North Koreans with their own MiG-15s, as well as training for their pilots. The results were remarkable and worrying. The Soviet air units claimed to have shot down 29 American aircraft through the month of November: 11 F-80s, 7 B-29s and 9 F-51s. Most of these claims tally with acknowledged losses, but US sources assert that most were either operational or due to ground-based anti-aircraft artillery.

The mission of the MiG-15 was primarily to counter the USAF Boeing B-29 Superfortress bombers and in this role it was highly successful. It wasn't long before the MiGs began to exact a heavy toll on the lumbering bombers and although one MiG-15 was shot down by an F-80C Shooting Star (piloted by

Lieutenant (Lt) Russell Brown in the world's first aerial engagement between opposing jet-fighters) there was nothing in theatre to counter the MiG. Losses among heavy bombers continued to rise and the Far East Air Force's High Command was forced to cancel the precision daylight attacks by B-29s and only undertake radar-directed night raids.

For the first time since early in the Second World War US forces did not have air superiority over a battlefield and urgent action was needed to regain this crucial advantage. The mission to get modern swept-wing fighters into the region to counter the MiG threat was given top priority by US military commanders and officials.

With the MiGs controlling the skies, the Defence Department ordered the 4th Fighter Interceptor Wing (FIW) to take its F-86A Sabres to Korea and regain air superiority and this unit was hurriedly directed to move to the combat theatre from New Castle County Airport, Delaware, on 11 November 1950. The introduction of the Sabre to the Korean War was a game changer. Although Soviet-piloted MiGs provided a more even match for the Sabres, Chinese and North Korean pilots were outclassed by their UN adversaries. Over the following two-and-a-half years of aerial combat there were to be several hundred MiGs shot down for a fraction of Sabres losses.

The 4th FIW's F-86As were ferried to either San Diego or McClellan Air Force Base. The Sabres at San Diego were loaded aboard the escort carrier USS *Cape Esperance* and those at McClellan departed from San Francisco

Project GUN-VAL

It didn't take long for the pilots in Korea to realise that the MiG-15 had more firepower than the F-86. Lt Col Bruce Hinton's briefing to commanders in January 1951 included the denunciation that 'the firepower of the F-86 is not sufficiently destructive, and should be modified with a calibre heavy enough to ensure structural damage with a minimum number of hits'. Reports from other pilots backed this up and also recommended that other improvements were needed, such as a more powerful engine and better gun sight – both of which were already being developed.

The Los Angeles Division of NAA was therefore contracted to modify twelve undelivered F-86Fs with rebuilt gun bays to accommodate four 20mm guns. Of these Sabres, ten were to receive Mauser guns designed in Germany during the Second World War but never installed in combat aircraft, and two were to receive Swiss-designed Oerlikon guns that had been successfully used in the Second World War. The Oerlikon installation eventually proved unsuitable; and after testing at Eglin Air Force Base, that portion of GUN-VAL was terminated.

The combat evaluation programme began in January 1953 and ran for 16 weeks, ending in April. GUN-VAL aircraft were often flown in mixed flights with 'ordinary' Sabres so as not to 'advertise' the presence of the cannon equipment. It has since been learned that Russian pilots who flew against the F-86 in Korea often derided the standard 0.50-calibre armament as 'pea-shooters', but once the 20mm cannons were fired, a MiG pilot soon realised that he was not up against a normal Sabre.

Although eight GUN-VAL aircraft arrived in the Far East in late 1952, one was held at Tsuiki Air Base, Japan, and remained there until late January when it was used to replace an aircraft lost due to a compressor stall and flame-out. It was discovered that this was caused by gun gas ingestion resulting in Capt Murray Winslow having to eject when he was unable to get an air-start. Winslow was rescued from the Yellow Sea. As a result the cannons' switches were modified to allow firing in pairs or all four together, with the pilots instructed to fire all four guns ONLY below 35,000ft, two between 35,000ft and 40,000ft and not to fire at all above 40,000ft.

During the 16-week trial period there were 6 confirmed MiG-15 kills and 3 probable. This system of 4 cannons later replaced the Sabre's 6 machine guns in the F-86H.

BELOW **Gun camera photograph of a MiG-15 being attacked by a USAF F-86.** *(USAF)*

aboard four tankers for shipment by sea. This was known as Operation Strawboss. The unit arrived at Yokosuka, Japan, in mid-December and established a rear echelon at Johnson Air Base on the outskirts of Tokyo. An advance echelon (Detachment 'A') was then deployed to Kimpo Air Base, South Korea, where operations commenced on 13 December 1950 with a series of familiarisation flights. Following this, two days later the 4th FIW opened its combat record with an uneventful orientation mission over North Korea.

The assigned mission was one of air superiority and this entailed the mounting of Combat Air Patrol (CAP) missions over north-

west Korea intended to prevent MiG-15s (flown from bases in Manchuria) from ranging further south. The rules of engagement permitted the 4th to attack and destroy the MiGs, but 'hot pursuit' into Chinese airspace was expressly forbidden and many MiGs were to escape as a result of this restriction. Despite these rules some F-86 pilots frequently initiated combat over MiG bases in the Manchurian

'sanctuary'. The hunting of MiGs in Manchuria would lead to many reels of gun camera footage being 'lost' if the reel showed the pilot had violated Chinese airspace.

The 4th FIW wasted no time in setting about its job and on 17 December Lt Col Bruce Hinton opened the account when he shot down a MiG-15 over the Yalu River.

For the best part of a year, the 4th FIW was

RIGHT This Sabre has suffered serious damage from a MiG-15's 37mm cannon. Although slow firing, the heavy-calibre weapon was devastatingly effective and was known on occasions to take an F-86's wing clean off. This particular aircraft was flown back to Johnson Air Base, Japan, where it was declared fit only for spare parts. (John Henderson)

the only Sabre-equipped unit in the war zone and it made good use of this exclusivity by accounting for most of the 102 MiGs claimed destroyed by the USAF during this period. In the autumn of 1951 the 4th began to convert to the much improved F-86E model. This change took several months because there was some urgency to replace the F-80Cs of the 51st FIW at Suwon Air Base with F-86Es,

thus some 'A' models remained in front-line service until the middle of 1952. The 51st FIW flew its first F-86E combat mission on 1 December 1951.

The 51st soon demonstrated the capabilities of the F-86E by accounting for 25 of the 31 confirmed victories in January 1952. Together with the 4th FIW they would account for a total of 375 MiG-15 'kills' in that year.

ABOVE At least 17 F-86Es of the 51st FIW on the Kimpo flight line being readied for a mission. *(USAF)*

Korea's poor runways

Jet aircraft, with their long take-off and landing runs, needed proper paved runways measuring 6,000ft or more. What they got in Korea was about 5,000ft of Pierced Steel Planking (PSP) laid on rough, rocky ground with enough lumps, bumps and uneven patches to ensure that a high proportion of jets suffered burst tyres on take-off or worse.

RIGHT F-86F sitting in a dispersal constructed from PSP (Pierced Steel Planking) and protected by sandbags. PSP was used extensively in Korea to create temporary runways and taxiways. This aircraft is loaded with HE 500lb bombs on the inboard pylons. An additional hard point was added to the 'F' to enable drop tanks and bombs to be carried simultaneously. *(USAF)*

Those first encounters established the main features of the aerial battles of the war. The MiG-15 and MiG-15bis had a higher ceiling than all the versions of the Sabre and accelerated faster due to their better thrust-to-weight. The MiG-15's climb rate was also greater than the F-86A and 'E' (later the F-86F would match the MiG-15's rate). A better turn radius at high altitude further distinguished the MiG-15 (all Sabres turned tighter below an altitude of 26,000ft) as did more powerful weaponry in the way of one 37mm N-37 cannon and two 23mm NR-23 cannons versus the inferior hitting power of the Sabre's machine guns. However, the MiG was slower at low altitude and the Soviet Second World War-era ASP-1N gyroscopic gun sight was less sophisticated than the accurate

A-1CM and A4 radar ranging sights of the F-86. Thus, if the MiG-15 forced the Sabre to fight in the vertical plane, or in the horizontal at altitude, it gained a significant advantage. Furthermore, a MiG-15 could easily escape from a Sabre by climbing to its ceiling, knowing that the F-86 could not follow. At low altitude the Sabre excelled over the MiG in almost every aspect excluding climb rate and was the victor most of the time, especially if the Soviet pilot made the mistake of fighting in the horizontal plane. Because of its extra weight and aerodynamic cleanness the Sabre could always out-accelerate the MiG in a dive.

The F-86E's all-moving tail-plane has been credited with giving the Sabre an important advantage over the MiG-15, although far greater

emphasis should be given to the training, aggressiveness and experience of the F-86 pilots.

A re-equipment programme was implemented in the summer of 1952 when the more powerful F-86F was introduced with the 51st FIW. Within three months the 'F' had also equipped the 4th FIW, with the aircraft being assigned to two more units during 1953. The first of these was the 18th FBW, which had previously flown the NAA F-51 Mustang, logging its last sortie with this Second World War veteran fighter-bomber on 23 January 1953.

With the introduction of the F-86F, the opposing aircraft were more closely matched, with many combat-experienced pilots claiming

ABOVE F-86E parked in sandbag revetments on an air base in Korea. The sandbags protected aircraft from bomb fragments as well as sniper fire from disgruntled locals. *(USAF)*

LEFT Armourers attend to a 51st FIW F-86E Sabre's 0.50-calibre M3 machine guns. This is an early 'E' fitted with the 'vee' type windshield.

BELOW RF-86F-30 photo-reconnaissance aircraft converted to that role in Korea. This photograph also shows the wing fence that identifies that the aircraft has the '6-3' wing fitted, which on this aircraft was factory fitted as opposed to an upgrade carried out in the field.

RIGHT The South
African Air Force's
No. 2 Squadron
transitioned from
F-51s to the F-86F
while attached to 18th
Fighter Bomber Wing.
(USAF)

BELOW Sqn Ldr
Eric G. Smith
(photographed not
climbing into his own
Sabre, but that of
1st Lt Earl Brown)
was one of several
British pilots to fly
Sabres during the
Korean War. Smith
was on an exchange
tour with the Royal
Canadian Air Force.
Canada had wanted
to deploy its own
fighter wing to Korea,
but instead provided
people and aircraft.
(Eric G. Smith)

a marginal superiority for the F-86F. In general
appearance and size the F-86A and F-86F were
much the same and therefore indistinguishable
to the enemy.

In addition to three USAF squadrons, the
18th FBW controlled the attached No. 2
Squadron, South African Air Force (SAAF), and
this unit also received F-86Fs, joining other
Wing elements in battle. No. 2 Squadron mainly
used the Sabre in the fighter-bomber role.

Within weeks of the 18th receiving the
F-86F, the 8th FBW began to follow suit, the
first squadron being withdrawn from combat

to enter training on 22 February. The other two
squadrons continued to use the F-80 for a little
while longer, until the 80th FBS recorded the
USAF's last Shooting Star operational sortie on
30 April 1952. The 8th FBW divided its assets
between MiG CAP and air-to-ground missions.

With many more Sabres available for
operations the enemy's MiG squadrons suffered
particularly heavy losses during the last seven
months of the war. No fewer than 287 were
claimed during this period. June 1953 was a
record-breaking month in almost every respect.
Some 77 MiGs were claimed and the sortie rate
peaked at 7,696. Allied attrition was higher too
with 14 Sabres being lost to enemy action and 9
more being written off due to operational causes.

Until the 8th and 18th FBWs converted
in 1953 the Sabre was only occasionally
employed as a fighter-bomber. Over 7,500 tons
of bombs and nearly 150 tons of napalm were
dropped along with 270 rockets fired against
ground targets. These are modest figures when
compared with the ordnance dropped by F-51s,
F-80s and F-84s, which bore the brunt of the
ground-attack mission. Nevertheless, the F-86
did make a valuable contribution in this role
during the conflict.

Hostilities in Korea ended on 27 July 1953.
In the three years of war as many US personnel
were killed in action as in the eleven years of
war in Vietnam. For the Sabre it was the end of
an extraordinary period in its history.

RIGHT **Lt Col Bruce Hinton was victorious in the first swept-wing versus swept-wing combat encounter.** *(Time)*

The first Sabre versus MiG-15 combat

Poor weather over Korea kept the Sabres on the ground until late morning on 17 December 1950. F-86A-5s from the 4th FIW flew their first combat patrol that day from Kimpo. Lt Col Bruce Hinton, the commander of the 336th Fighter Interceptor Squadron and a veteran pilot, led a flight of four F-86s north to patrol MiG Alley near Sinuiju, across the Yalu River from Antung, China. Col Hinton used F-80 radio call signs, flying patterns and altitudes to deceive and entice the MiGs into combat. The trick worked. While the Sabres patrolled in a finger four formation at 25,000ft, four MiGs took off from Antung to attack. The MiGs approached at 18,000ft, climbing in a south-easterly direction. Col Hinton radioed the other Sabres to release their drop tanks and dived on the MiGs as they paralleled the Yalu River.

As Col Hinton banked to close behind the MiGs, he became separated from his flight. The Sabres closed at 410 knots and 20,000ft while increasing their airspeed at maximum power. The MiGs dived and then began a climbing left turn. Col Hinton's airspeed on his early F-86A exceeded the red line on his Mach meter at this point. Then the MiGs broke. Col Hinton fired a long burst from his machine guns, hitting the second MiG-15 in its right wing and fuselage, which began smoking. Col Hinton then fired a long burst and saw fire exit the MiG's tailpipe. He fired another long burst, and fire covered the entire rear of the MiG's fuselage. As the MiG slowed, Col Hinton opened his dive-brakes and throttled the engine back. The MiG was given another long burst. Pieces flew off as the MiG rolled on its back and went down. The stricken MiG-15 crashed 10 miles south-east of the Yalu River. Col Hinton fired 1,200 rounds on his attack.

One other MiG was damaged by Capt Morris Pitts, but all three remaining MiGs outran the Sabres across the Yalu River to Manchuria. Col Hinton later flew a victory roll as he returned to Kimpo.

Korean War – Sabre 'aces', claims, victories and losses

A fighter 'ace' is a pilot credited with shooting down five enemy aircraft during aerial combat. Of the forty-one UN pilots who earned the designation of 'ace' during the Korean War, all but one flew the F-86 Sabre

LEFT **Forty USAF F-86 pilots (and one Marine Corps exchange pilot) became Korean War aces. Not pictured is 1st Lt (later Lt Gen) Charles Cleveland, who belatedly received credit for his fifth kill in 2008.** *(USAF)*

(the exception being a Navy F4U Corsair night-fighter pilot). Some were double 'aces' and at least one was a triple 'ace'.

The 'ace' pilots are acknowledged to have accounted for the impressive tally of 305 enemy aircraft, almost all of which were MiG-15s; close to 500 other MiGs were claimed by non-'ace' Sabre pilots during more than 87,000 F-86 combat sorties logged by UN forces during the war. In amassing that extraordinary record, 110 F-86s are known to have fallen victim to enemy action, MiGs definitely accounting for 78 of them, while 19 were claimed by ground fire and 13 were lost to unknown causes. Other forms of attrition claimed 114 aircraft, made up of 61 to operational causes, 13 recorded as 'missing' (a figure that may well include some MiG victories) and 40 to non-operational accidents, 6 of which happened at bases in Japan.

Of course there is a caveat to all of this. When looking at the tally for victories and losses during the Korean War more broadly the 'official' figures need to be treated with some caution. It must be remembered that the history of any conflict is usually written by the victor, thus it is the UN side that is told and accepted as the truth although there is almost certainly another side to the story. This doesn't mean that the 'kill' ratios commonly quoted are inevitably wrong; it just means that there needs to be some qualifications made and provisos considered. For example, what qualifies as an aerial combat victory? It is known that some

'kills' were awarded to US pilots on the strength of gun-camera film only; however, it does not necessarily follow that images of MiGs being hit by the Sabre's 0.5in guns resulted in the aircraft crashing. Indeed, it is known that the MiG could soak up a great deal of punishment and many were able to return to their bases, albeit somewhat battle-scarred. Because the MiG was armed with cannon it was much rarer for Sabres to limp home if similarly caught; a direct hit with cannon was usually final and fatal. It also has to be appreciated that any MiG kills north of the Yalu River were discounted by the UN forces and hence there were a number of these 'illegal kills' not recorded.

Soviet pilots are known to have fared much better than their Korean or Chinese comrades. Russian pilots were mostly experienced veterans from the Second World War and their participation in the war was highly classified. The Soviet Air Force supplied at least three air regiments to support the North Korean Air Force. Based north of the Yalu River in mainland China, the Soviet units included many veteran fliers from the Second World War. One of them, Maj Dimitrii Yermakov, was a Second World War multiple 'ace' with 26 victories.

Whereas 'kill' ratios in the order of 10:1 in favour of the UN are commonly suggested it is known that the score was much more evenly matched in contests between UN and Soviet pilots. According to former Communist sources, Soviets initially piloted the majority of MiG-15s that fought in Korea. By early 1953, once enough Chinese and North Korean pilots had been trained, most Soviet pilots were withdrawn, but not before losing a great number of experienced aviators. By their own admission, the Russians lost four Soviet-piloted MiGs to every F-86 shot down. Their archives also note that Chinese MiGs went down at a ratio of almost 8:1. No figures have come from this source regarding North Korean MiG losses.

It is also known that losses due to enemy action were not always recorded properly by the UN. For example, an aircraft that ran out of fuel returning from a mission (because, for example, a cannon shell had shot a hole in a fuel tank) was sometimes chalked down as an operational (not combat) loss. It is generally accepted that US claims erred on the generous side with

BELOW RF-86A 'Honeybucket' (48-217), which was lost on a later mission, seen in 1952. Seven F-86As were converted to a reconnaissance configuration in Korea: the in-theatre modification added a forward oblique 24in K-11 camera and two 20in K-24 cameras mounted lengthways with a mirror arrangement to provide vertical coverage. These aircraft flew dangerous unescorted missions over MiG Alley in what was known as Operation Honeybucket.

First-hand account of combat with the MiG-15

Maj Winton 'Bones' Marshall, CO of the 335th FIS, shot down three Soviet- and two Chinese-piloted MiG-15s in Korea. His most difficult and remarkable success came in a melee which began at 8:50 hrs on 28 November 1951, and lasted for 20 minutes. Maj Marshall recalled:

I'll remember this mission for all the days of my life. Never have I had to fight so hard to survive. We had just rolled out from breaking into the two attacking MiGs, when almost immediately, my cockpit was surrounded by a hail of bright tracers from the cannons of the MiG on my tail. They looked the size of oranges as they went by the canopy. My wingman had said nothing. I jerked the stick back so hard that I easily exceeded the aircraft G limits. The MiG went one direction and my wingman the other. I had lost them both.

Before I could take a breath, I was again the target of a stream of tracers, but a lot closer this time. There was a second MiG sitting right on my tail. Startled, I again slammed the stick back, trying to 'split S' out of there. It produced spectacular results. My Sabre did a neat snap roll, and I ended up in an inverted spin with zero airspeed. This was a great evasive manoeuvre, I thought. No one could have stayed with me through that gyration.

Except, when I looked out through my canopy, there was a MiG. And we were both spinning down together, canopy to canopy. In seconds I made a quick spin recovery. But so did he. We had ended up in a flat spin, with very little airspeed, in a nose-up attitude. Except that his nose was almost pointing at me. I expected him to start firing at any moment.

As an illustration of the flight stability of the F-86, my aircraft completely responded when I again slammed the sloppy stick to one side, kicking the rudders as hard as I could. It worked! I was off in another spin, slower but more controllable this time. Then the impossible happened. There was my MiG, also spinning down in the same air space with me. We were fast losing too much altitude,

LEFT Maj Winton 'Bones' Marshall.

so I again made a spin recovery. The MiG recovered right beside me. Except this time, he was at my 12 o'clock position – directly in front of me. It was a simple task to open fire. I must have hit something vital as the MiG suddenly caught fire and exploded.

What a great fighter pilot that MiG guy was, I thought he was Ol' Casey Jones* himself. Other Sabre pilots that had seen the fight, said that the MiG appeared to have stuck with me in my hard break, until we both snap-rolled and fell off in that first spin. That MiG driver proved to be one hell of a wingman, when he stuck with me in a formation of sorts, as we both spun down the second time, with destination unknown. It was just Lady Luck riding with me, that he ended up in the dead centre of my gun sight. Otherwise, I don't know. I was well below 'Bingo' fuel, and it was either then or I wouldn't have made it home.

* 'Casey Jones' wasn't a specific person, but an airman who played the role of squadron leader. Russian loss records confirm that the identity of 'Casey Jones' that day was 1st Lt German Timofeyevich Shatalov, a pilot of the elite 523rd IAP/303rd IAD, a very skilful MiG-15 pilot credited with five aerial victories against US aircraft.

LEFT Portrait of Starshii Leitenant (1st Lieutenant) German T. Shatalov, the skilful MiG-15 ace shot down and killed by Winton 'Bones' Marshall on 28 November 1951.

Sabre scorecard

The following figures are 'official' and based on the *USAF Statistical Digest Fiscal Year 1953* covering the period from December 1950 to July 1953 inclusive.

RIGHT Smoke pours from a MIG-15 as bullets spatter from the blazing machine guns of a USAF F-86.

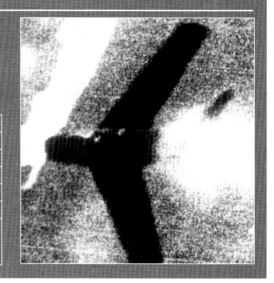

Average number of Sabres in theatre	184
Total number of sorties flown	87,177
Number of MiGs claimed shot down	792
Other enemy aircraft claimed	22 (including 4 on ground)
Enemy aircraft probably destroyed	119
Enemy aircraft damaged	818
Sabres lost to enemy fire	110 (78 in air-to-air combat, 19 to ground fire, 13 to causes unknown)

BELOW MiG-15 No. 346 at Bornholm, Denmark. This aircraft was the first example to fall into allied hands by defection and was extensively 'examined' before being returned to Poland.

some observers being prepared to argue for an 'over-claim' factor of as much as 50 per cent. Of course the North Korean authorities made extravagant claims too, which unbelievably had the number of Sabre losses exceeding (by quite a margin) the number of aircraft that were sent to the theatre. In other words, it is difficult to be completely certain about the claims made by either side concerning losses and victories during this war.

Defection

Eager to obtain an intact MiG for testing, the USA offered a reward of US$100,000 and political asylum to any pilot who would defect with his MiG-15. Franciszek Jarecki, a pilot of the Polish Air Force, defected from Soviet-controlled Poland in a MiG-15 on the morning of 5 March 1953, allowing Western experts to examine the aircraft for the first time. Jarecki

flew from Słupsk to the field airport at Rønne on the Danish island of Bornholm. The whole trip took him only a few minutes. There, specialists from the USA, called by Danish authorities, thoroughly checked the plane. According to international regulations, they returned it by ship to Poland a few weeks later. Jarecki eventually became a US citizen.

Others began to follow his example, such as the North Korean pilot 21-year-old Senior Lt

No Kum-Sok, who claimed to be unaware of the US$100,000 reward when he landed at Kimpo Air Base on 21 September 1953. Shortly after landing the young pilot not only learned of the reward but also that his mother had been safely evacuated from North to South Korea in 1951 and that she was still alive and well.

Kum-Sok's MiG-15 was taken to Okinawa where it was first flown by Wright Field test pilot

Capt Tom Collins. Subsequent test flights were made by Capt Collins and Maj Chuck Yeager, who reported in his autobiography that the MiG-15 had some dangerous handling faults. Soviet pilots who learned that Yeager had flown the MiG in a high-speed dive were incredulous as this was supposedly very hazardous and something they were forbidden from doing! In fact, although the MiG-15 did have some handling quirks and could, in principle, exceed flight limits in a dive, its air-brakes opened automatically at the red-line limit, preventing it from going out of control.

The aeroplane was next disassembled and airlifted to Wright-Patterson Air Force Base, Ohio, in December 1953 where it was reassembled and given exhaustive flight testing. The USA then offered to return the aeroplane to its 'rightful owners'. The offer was ignored, and in November 1957 it was transferred to the National Museum of the United States Air Force near Dayton, Ohio, for public exhibition, where it carries its original number '2057'.

At his request, No came to the USA, changed his name and became a US citizen. He graduated from the University of Delaware, was joined by his mother and was later married.

Interestingly, just below the gun sight on Lt No's MiG-15 was the following admonition in red Korean characters: 'Pour out and zero in this vindictive ammunition to the damn Yankees'.

Soviet Sabre

During the Korean War, the Soviets were also looking out for an intact US F-86 Sabre for evaluation/study purposes. Their search was frustrated, largely due to the US military's policy of destroying their weapons and equipment once they had been disabled or abandoned; and in the case of US aircraft, USAF pilots destroyed most of their downed Sabres by strafing or bombing them. However, on one occasion an F-86 was downed in the tidal area of a beach and subsequently submerged, preventing its destruction. The aircraft was dismantled and taken to Moscow and a new OKB (Experimental Design Bureau) was established to study it, which later became part of the Sukhoi OKB. The F-86 studies contributed to the development of aircraft aluminum alloys, but it is believed that the Soviets learned more from the aircraft's systems.

First Taiwan Straits Crisis – 1955

The first Taiwan Straits Crisis was the first combat action to be flown by Sabres since the end of the Korean War and was sparked off in 1954 by Chinese attacks on Nationalist-held islands. The USAF deployed Sabres to theatre in early 1955 and the RoCAF received the first deliveries of their own aircraft during this period. The crisis had ended by May 1955, but sporadic combat between the People's Liberation Army Air Force and Republic of China Air Force F-86Fs continued over the next few years.

Second Taiwan Straits Crisis – 1958

Further combat between the Sabres and MiGs came in August and September 1958, when Communist China (PRC) tried to impose a blockade on the islands of Quemoy (which lies about 10km from mainland China) and Matsu, which were occupied by Nationalist Chinese forces (ROC). This was accompanied by an aggressive propaganda assault on the USA, threats against US naval ships and a declaration of intent to 'liberate' Taiwan. The PRC did not carry out any attack or attempt an invasion of the Taiwanese main island, and thereby did not deliberately provoke a war with the Americans. This was just as well given the US administration's view that any attack on Taiwan by the PRC could ultimately lead to retaliation by US forces in the area, with a high possibility of nuclear weapons being employed.

Previously, in the spring of 1955, President Eisenhower had sent a mission to persuade Chiang Kai-shek, the ROC president, to withdraw from Quemoy and Matsu islands because the Americans believed that they were exposed to just such an attack. Chiang Kai-shek would not withdraw and subsequently Eisenhower was compelled to provide the Nationalists with military assistance, which included howitzers capable of firing nuclear shells. Then in 1957 an agreement between the US and the ROC placed Matador missiles on Taiwan and these surface-to-surface weapons (capable of carrying conventional or nuclear warheads up to 600 miles) unquestionably led the PRC to view the situation in the strait as hostile and somewhat menacing and it carried out artillery attacks on the islands.

Once the shelling began, the USA made it clear that it would support the ROC in the defence of the islands when Secretary of State John Foster Dulles declared that the USA would take 'timely and effective action to defend Taiwan'. As tension mounted between the USA and China, the US Joint Chiefs of Staff developed plans for nuclear strikes on the Chinese cities of Shanghai, Guangzhou and Nanjing. These plans were consistent with the public statements by Dulles, who on 12 January 1954 had threatened 'massive retaliation' against Communist aggression and expressed willingness to go 'to the brink' of war to stop such aggression.

Despite Soviet support of the PRC's claims to the islands, the bombardment abated, then virtually ceased after President Eisenhower warned that the USA would not retreat 'in the face of armed aggression'. The unexpectedly forceful US response surprised Chinese and Soviet leaders, and on 6 September 1958 Zhou Enlai proposed a resumption of ambassadorial-level talks with the USA in order to reach a conclusion to the crisis. The crisis officially ended on 6 October 1958 when Chinese Minister of National Defence Marshal Peng Dehuai offered to negotiate a peaceful settlement with the Nationalists and announced that the PRC would suspend the bombardment for one week.

Under a secret effort known as Operation Black Magic, the US Navy modified some ROC Air Force F-86 Sabres with its newly introduced AIM-9 heat-seeking Sidewinder air-to-air missile to provide an edge against more advanced PRC MiG fighters, which had an altitude advantage over the Sabre. This weapon fit proved reasonably successful in accounting for a number of MiG-15s and MiG-17 Frescos. On 24 September 1958, the Sidewinder was used for the first time in dogfights with 32 Sabres against over 100 MiGs; 10 MiGs were downed by the F-86s in a single day.

Specific details about encounters during the short conflict are hard to verify. Nationalist forces claimed the destruction of twenty-nine MiGs at a cost of just two F-84G Thunderjets, however it seems probable that these claims were overgenerous. More certain is the view that, as in Korea, pilot skill and superior handling qualities proved to be decisive.

ABOVE A flight of four Pakistani Air Force Sabres. The F-86 was the most numerous fighter deployed during the Indo-Pakistani War of 1965.

Indo-Pakistani War – 1965

The Indo-Pakistani War between April and September 1965 was a culmination of skirmishes that took place between India and Pakistan over the disputed region of Kashmir. Previously the two countries had clashed over the territory in 1947.

Much of the war was fought by the countries' land forces in Kashmir and along the International Border between India and Pakistan. This war saw the largest amassing of troops in Kashmir since the Partition of British India in 1947. Most of the battles were fought by opposing infantry and armoured units, with substantial backing from air forces and naval operations.

RIGHT Sqn Ldr Mohammed Mahmood Alam, the most successful pilot of the Pakistani Air Force during the Indo-Pakistani War of 1965.

Pakistan's exploits with the Sabre in battle were by no means as one-sided as those of the UN allies in Korea or Nationalist China, but they did nevertheless add further to an already impressive career and there were one or two outstanding moments in both the air-to-air and air-to-ground arenas.

Foremost among these were the actions of Sqn Ldr Mohammed Mahmood Alam. On 7 September 1965 Alam encountered a clutch of Indian Hunters about 30 miles (48.3km) to the east of the main Pakistani fighter base at Sargodha. Within a few minutes he had shot down four of the enemy to add to another victim claimed minutes earlier, thereby technically achieving the much-vaunted 'ace' status in a single sortie. He had, in fact, opened his account on the previous day when he destroyed two Hunters that intervened to prevent a strike on the Indian air base at Adampur. Alam subsequently accounted for two more Hunters in another engagement on 16 September, ending the short war as the top scorer in the Pakistan Air Force (PAF), with nine confirmed 'kills'. Eight of those were scored with the F-86F's guns, proving that this tried and tested weapon was still a valid element of fighter aircraft armament. Alam's ninth and last victim fell to an AIM-9B Sidewinder. This missile was a weapon far from being as lethal as it is today, but was nonetheless treated with great respect and for which the Indian Air Force (IAF) had no direct equivalent. Since Indian losses in air combat were eventually held to be about 22, Alam was personally responsible for almost half.

Other Sabre pilots added to the list and the F-86 ended up victorious over about 12 Hunters, a couple of Gnats and 4 de Havilland Vampires. On the debit side Pakistan

BELOW On 6 September 1965 eight F-86Fs of No. 19 Squadron led by Sqn Ldr Sajjad Haider struck the Indian Air Force base at Pathankot. With carefully positioned dives and selecting each individual aircraft in their protected pens for their strafing attacks, the strike elements completed a textbook operation. Wg Cdr M.G. Tawab, flying one of the two Sabres as tied escorts overhead, counted fourteen wrecks burning on the airfield. Among the aircraft destroyed on the ground were nearly all of the Indian Air Force's Soviet-supplied MiG-21s.

LEFT Sqn Ldr 'Nosey' Haider who led the Pakistan Air Force attack on Pathankot Air Base.

acknowledged the loss of seven Sabres in air combat to the diminutive Folland Gnat, which proved to be fast, nimble and hard to see. The IAF Gnats earned the nickname 'Sabre Slayer'.

The PAF Sabres also performed well in ground-attack missions with claims of destroying around 36 aircraft on the ground at Indian airfields at Halwara, Kalaikunda, Baghdogra, Srinagar and Pathankot. India only acknowledges 22 aircraft lost on the ground to strikes attributed to the PAF's F-86s and its B-57 Canberra bombers.

Bangladesh Liberation War – 1971

In later years, the Sabre returned to battle on the side of Pakistan in the Indo-Pakistani War of 1971. The war was precipitated by the crisis brewing in East Pakistan. Following Operation Searchlight and the 1971 Bangladesh

BELOW After debriefing and interrogation, the Pathankot strike element were credited with seven MiG-21s, five Mysteres and one Fairchild C-119 destroyed on the ground, plus damage to the air traffic control building. Later assessments by the Pakistan Air Force Director of Plans and Operations were inclined to consider the number of MiGs credited as slightly optimistic, while the Indians claimed they were not MiG-21s at all, but Mystères. Officially, India accepted losing ten aircraft during this strike, including one MiG-21. The Indian Air Force official history states that only nine MiG-21s were in service at this time, and India was prepared to show eight at the end of the campaign. But whatever the precise type of destruction caused at Pathankot, it was an undisputed success in being inflicted without loss.

atrocities, about 10 million Bengalis in East Pakistan took refuge in neighbouring India. Because of the impending humanitarian crisis and its own interest in dismembering Pakistan, India intervened on the side of the Bangladesh liberation movement. Within two weeks of intense fighting Pakistani forces surrendered to India, following which the People's Republic of Bangladesh was created. This war saw the highest number of casualties in any of the India–Pakistan conflicts, as well as the largest number of prisoners of war since the Second World War after the surrender of nearly 90,000 Pakistani police and civilians.

The Canadair Sabres, along with some older PAF F-86Fs, were the mainstay of the PAF's day-fighter operations during the Bangladesh Liberation War. Newer fighter types such as the Mirage III and the Shenyang F-6 had been acquired but were available in only small numbers.

In East Pakistan only one Sabre Mk 6 squadron (14 Squadron) was deployed by the PAF to face the numerical superiority of the formidable IAF Soviet MiG-21 and the Sukhoi SU-7 fighters.

Despite these odds, the PAF F-86s performed well, with claims of 31 Indian aircraft shot down in air-to-air combat, including 17 Hawker Hunters, 8 Sukhoi Su-7 'Fitters', 1 MiG-21 and 3 Gnats, all for the loss of 7 F-86s. India, however, claims to have shot down 11 PAF Sabres for the loss of 11 combat aircraft. The IAF numerical superiority overwhelmed the sole East Pakistan Sabre squadron, enabling complete air superiority for the IAF.

In East Pakistan 11 Sabres were disabled by PAF forces to keep them out of enemy hands, but five were recovered in working condition and flown again by the Bangladesh Air Force.

After this war, Pakistan slowly phased out its F-86 Sabres and replaced them with Chinese F-6 (Russian MiG-19-based) fighters. The last of the Sabres were withdrawn from PAF service in 1980.

Guinea Bissau – 1961–64

The African Party for the Independence of Guinea and Cape Verde (PAIGC) was founded in 1956 by Amilcar Cabral with the goal of negotiating independence for Portuguese Guinea and the Cape Verde Islands from their Portuguese colonisers. When the negotiations failed, the PAIGC worked to gain the support of the Guinean villagers and in 1961 guerrilla bands began attacking Portuguese army posts and police stations. With the guerrillas entrenched in the jungles, the Portuguese government responded by deploying units to AB.2 at Bissau/Bissalanca to reinforce NAA T-6G Texans already stationed there. It also increased troop numbers (eventually growing to about 35,000) and began to bomb and raid guerrilla hide-outs.

At first eight F-86Fs were deployed in Guinea at Bissau/Bissalanca Air Base and used in ground-attack and close-support operations. These aircraft formed 'Detachment 52' of the Esquadra 51, based at the Base Aérea 5, in Monte Real, Portugal. During this action one aircraft was shot down by enemy ground fire on 31 May 1963; the pilot ejected safely and was recovered. Several other aircraft suffered combat damage, but were repaired and one was written-off in a landing accident.

By early 1964, the Sabres of Detachment 52 were in very poor condition and intense pressure from the USA caused the Portuguese Air Force to cease their operations, deploying T-6s instead. The Sabres were returned to the Portuguese mainland, having flown 577 combat sorties, of which 430 were ground-attack and close air support missions.

As PAIGC began to gain more control over the country they established civilian rule and held elections for a national assembly. By 1973 the PAIGC gained control of roughly two-thirds of Portuguese Guinea and declared independence over their portion of the Portuguese colony. The new country was named the Republic of Guinea-Bissau. Portugal initially refused to recognise it, although following the Portuguese military coup in 1974, Portugal finally granted independence to Guinea-Bissau.

The Sabre saw a limited amount of action in service with air arms in Latin America and in other minor conflicts, but the F-86 never again enjoyed such a clear-cut margin of superiority as was demonstrated in Korea.

Chapter Three

Sabre 48-178

The world's only airworthy F-86A

───●───

The F-86A featured throughout this book, USAF serial number 48-178, was manufactured at NAA's Inglewood factory in Los Angeles, California, and became available to the Air Force on 23 February 1949. The aircraft was the 39th F-86A-5 sub-type to be built (the 72nd Sabre built). The 'dash-5' was the first production-standard model powered by a 5,340lb thrust General Electric J47-GE-7 turbo-jet.

OPPOSITE Pat Palmer looks over No. 178, which lacked an engine and had no wing leading edges. *(Jim Larsen)*

The original logbook for 48-178 shows that it was available on 23 February, accepted on 14 April and delivered to the USAF on 18 April 1949. (Via AFHSO research)

ABOVE The original logbook for 48-178 shows that it was available on 23 February, accepted on 14 April and delivered to the USAF on 18 April 1949. *(Via AFHSO research)*

ABOVE Insignia of the 94th Fighter-Interceptor Squadron, the first unit to which 48-178 was assigned.

RIGHT Sabre 48-178 in the colours of the units she served with during her USAF and Air National Guard career. *(© Duncan Curtis)*

F-86A s/n 48-178
94th FIS, March AFB, California
1949

F-86A-5 s/n 48-178
116th FIS Washington ANG, Moses Lake AFB, Washington
1951

F-86A-5 s/n 48-178
93rd FIS, Kirtland AFB, New Mexico
1952

F-86A-7 s/n 48-178
469th FIS, McGhee-Tyson AFB, Tennessee
1953

F-86A-7 s/n 48-178
196th FBS, California ANG, Ontario ANG Base
1955

F-86A-7 s/n 48-178
196th FBS, California ANG, Ontario ANG Base
1958

ABOVE General Robin Olds experienced the Sabre for the first time in 48-178. Olds was a fighter pilot of great distinction and a 'triple ace', with a combined total of 16 victories in the Second World War and the Vietnam War. This photograph was taken during the Vietnam War.

ABOVE The wife and young son of crew chief Harold Shaw pose in front of No. 178 at March Air Force Base in February 1950. Harold was personally in charge of this aircraft for over a year.

ABOVE The starboard machine-gun servicing panel has been removed in this photograph of 48-178 while in service with the California Air National Guard and before the serial number was changed from four to five digits.

LEFT A line-up of 469th FIS Sabres: 48-178 is visible about halfway down the line.

LEFT Based at Ontario, California, 48-178 (by then converted to -7 standard) leads a formation of four F-86A-7s during its service with the California Air National Guard (48-178 leading; 49-1122 as No. 2; 49-1217 as No. 3; and 49-1336 flying as No. 4). This photograph was taken between May 1954 and April 1957 after which the 'buzz' number was changed from 8178 to 80178. (via Jim Larsen)

LEFT 48-178 was assigned to the 196th Fighter Bomber Squadron of the California ANG at Ontario in 1954. Maj Archie Nogle flew the aircraft on her final Air National Guard flight on 22 September 1958.

RIGHT As discovered by Jim Larsen in 1970, the remains of 48-178 were languishing with a scrap dealer in Fresno, California, who was asking $700 for the aircraft. *(Jim Larsen)*

BELOW At the end of its service 48-178 was donated to a vocational college in Fresno as an instructional airframe. It was later sold to raise funds and towed across the road to the reclamation yard where it was discovered. This explains why the airframe survived being 'chopped' and put beyond use. *(Jim Larsen)*

The restoration of 48-178

After being withdrawn from service 48-178 would have eventually followed many other Sabres to the smelters had it not been for Jim Larsen, a photographer, engineer and enthusiast, based in Seattle, Washington. Jim found out about the aircraft from a friend who, while on a business trip to California, had discovered the F-86A languishing in the yard of a scrap-metal dealer in Fresno. He had taken a few Polaroid pictures of the aircraft and noted the telephone number of the dealer on the back. The asking price for the aircraft was $700.

11 years of neglect had not been kind: the cockpit was stripped bare, numerous parts were missing and there was no engine.

US government policy at the time was that no US combat aircraft would be allowed to pass from Federal inventory to private concerns without being demilitarised. This usually meant cutting through wing spars and other specified structural points to the degree that return to flight status would be impossible. The days of purchasing obsolete fighter aircraft, such as was enjoyed after the Second World War, were gone forever. As it was, the purchasing of aircraft from foreign government disposals offered the

LEFT The demilitarised remains of F-86A 47-606, the second production F-86A-1 to roll off the assembly line, parts of which still fly today in 48-178, is removed from a scrapyard in preparation for the long road journey to Seattle. Note the wings have been 'chopped' near to the wing root. *(Jim Larsen)*

only potential for flyable jet aircraft. However, in the case of 48-178, a restoration project was possible because the basic structure was still intact and she had not yet been put beyond use.

It was clear that the undertaking would require lots of work (and money) and Jim enlisted the help of his good friend and former T-6 and P-51 owner/pilot Ben Hall, also of Seattle. Ben and Jim flew to Fresno where, although they were shocked at the condition of the aircraft, they assessed that rebuilding the Sabre was feasible. And so 48-178 was purchased by Jim for $500 (later sold on to Ben Hall) with the intention of a rebuild to flying condition. The owner of the Fresno scrapyard told Jim and Ben of another aircraft sitting at a nearby dealership and a deal was done for that aircraft too. This 'spares' airframe was the demilitarised remains of F-86A 47-606. Both airframes were loaded on to a truck and carted 1,000 miles north to Seattle where the restoration work was to be undertaken.

The risks were considerable in taking on this project and it would be four long and difficult years before they would know for certain that their initial judgement had been correct. The early restoration team was made up of Ben Hall, Rad Kostelnik, Howard

RIGHT When discovered, the cockpit of No. 178 had been stripped bare and was a mess. Note the armour plating that folds up behind the pilot's head at the rear of where the ejection seat should be is still present. *(Jim Larsen)*

ABOVE Ben Hall inspects the basic structure of 48-178. Although largely intact, it was clear that the undertaking to restore the aircraft was colossal. *(Jim Larsen)*

Mercer and Roy Lund. Additional help and assistance would also be forthcoming from former engineers from NAA, GE and Chandler Evans (the manufacturers of the fuel control system). A number of surplus sales dealers in California, Washington and Arizona also spent considerable time and effort attempting to locate various elusive parts and components.

As purchased, 48-178 had many shortcomings. Most notably it lacked an engine, had no wing leading edges and the cockpit was a mess. On the upside the canopy was in exceptional condition (it flies on the aircraft today) and there were boxes and boxes of spare parts. The second airframe provided most of the missing parts needed, including much of the cockpit; however, there were no usable leading edges. Various leading edges were acquired, although this aspect of the rebuild was to

present a seemingly endless set of problems. The design of the slat system changed several times during the production run of the 'A' model and unfortunately the slats obtained were incompatible with the cut-outs in the wing. It was decided that the best solution was to install the leading edges in the fixed closed position. To compensate for the lack of a leading edge device, a wing fence was attached about two-thirds of the way down the span of the wing. This resulted in an F-86A with a wing that resembled the F-86F '6-3' hard wing, although without the root and tip chord extensions of that modification.

One of the most elusive of all missing parts was an aileron control bracket, an intricate casting that would have been impossible to duplicate. It took nearly two years for the part to be located.

A serviceable tailpipe was another component that was very difficult to find. At first a few used pipes were fitted, but one of these actually blew up on the initial engine run-up and another failed when the thermocouple boss was blown out. Fortunately, McClellan Air Force Base was selling surplus examples and Ben was able to buy 11 assemblies, 5 of which were brand new.

Progress toward getting the engine airworthy was helped by Bill Heine joining the team as power plant specialist. Several J47 engines were located and purchased for prices ranging from as little as $50 up to $1,500, including a zero-time example for $350. At first efforts to get the engine running properly were thwarted by the simple mistake of fitting the emergency fuel switch in the cockpit upside

RIGHT Prominent 0.5in machine-gun ports: the business end of the Sabre. The guns were staggered so that the feeds from the ammunition trays did not need to bend by more than a few degrees. This is 47-606, which was used for parts in the restoration of 48-178. *(Andrew Critchell, aviationphoto.co.uk)*

down. Eventually this problem was identified, but only after extensive investigation and much frustration. Later in the restoration the engine would once again cause a major setback when it was started with the engine blank in place. Although the cover became wedged ahead of the compressor, the steel retaining pin attached to the 'Remove Before Flight' ribbon was ingested and the brand-new engine was destroyed. Ben would later admit that this incident was the lowest moment of the entire project and one of the few times in his adult life that he had broken down and cried. The project stalled again for several months while parts from two engines previously purchased for $50 each were combined to provide an airworthy engine.

Additional major restoration sub-projects involved hydraulics (for which 90% of the system was missing), a complete electrical rewiring job and locating and installing a new fuel control system.

Progress at times was painfully slow: the project was a weekend and evening effort. Furthermore, tragedy struck in September 1971 when two of the rebuild team (Rad and Howard) were killed in a T-6 crash on a flight back from participating at the Reno Air Races. The shock of this loss halted the project until the spring of 1972 when Ben decided to devote himself to the project full time. Ben's commitment was not only in time and money. He sold his business partnership and T-6 and yet still needed to seek additional financing to finish the project.

Along the way, there were many, many other problems to be solved and the team benefited

at times from outside help such as famed NAA chief test pilot Bob Hoover, who helped locate NAA engineers and documents to answer questions.

At the time that Ben acquired 48-178 there were no American-built jet-fighters flying in civilian hands (there were some foreign-built Vampires, T-33s and Canadian Sabres) and therefore, as far as the Federal Aviation Administration was concerned, there was no precedent for registering such an aircraft. The bureaucratic hoops that Ben had to jump through to obtain registration would

have been comical had they not been so time consuming, costly and frustrating. It eventually took nearly two years for registration N68388 to be officially bestowed upon the machine.

On 26 January 1974 the first series of taxi tests were successfully completed and then eventually on 24 February, after 4 years, 10,000 hours and immeasurable blood, sweat and tears, the aircraft flew from Paine Field, Washington, on a 30-minute flight over Puget Sound. Ben put aside his own personal desire to fly the aircraft in favour of a more current and experienced pilot. And so it was that Paul Bennett, a Boeing test pilot then flying a Sabre Mk 5 chase plane for the company, would take No. 178 into the air for its second 'first flight'. The aircraft flew with virtually no snags. Two weeks later a second flight, again with Paul Bennett at the controls, saw the aircraft achieve 41,000ft and a dive to Mach .95. Subsequent flights confirmed that the project had been a resounding success.

After checking out in a T-33, Ben Hall first flew the Sabre on 3 May 1974. The first public display followed three months later when the legendary Bob Hoover flew a spectacular aerobatic sequence at the Abbotsford International Air Show in British Columbia, Canada.

Ben maintained and flew No. 178 for thirteen years; however, he gradually found the Sabre to be more and more expensive to operate, and he finally sold it to John Dilley in Fort Wayne, Indiana, during 1988. Prior to this, in October 1983 the aircraft was re-registered N178 and was painted in Imron silver as a Korean War 4th FIW machine, although with incorrectly coloured (yellow and

LEFT No. 178 undergoes a repaint from the original post-restoration dark slate-grey scheme to white with orange and red stripes. *(Jim Larsen)*

black, which should have been white and black) identification stripes. It is thought that these stripes were mistakenly applied due to an erroneous colour scheme featured in a *Profile Publications* F-86A pamphlet from 1965.

Ben described the project as one of the great loves of his life and in 1992 (after it had been sold on to the Golden Apple Trust in the UK) 48-178 was nominated for the Rolls-Royce/Warbirds Worldwide Award for best jet restoration and voted the winner. This was a fantastic accolade and recognition of

ABOVE The restoration team comprising Ben Hall, Bill Heine and Roy Lund with newly restored 48-178. *(Jim Larsen)*

LEFT AND BELOW In what might be termed air-racer style, the aircraft was repainted white with a red/orange/yellow line, nose to tail. *(Jim Larsen)*

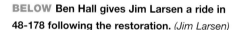

RIGHT In 1989 Bill Hall sold N68388 to John Dilley in Fort Wayne, Indiana.

BELOW In 1983 the aircraft was re-registered as N178 and repainted as an early Korean War 4th FIW machine, albeit with incorrect yellow and black identification stripes. This photograph was taken at Fort Wayne International Airport (previously Baer Field military base). *(Gary Chambers)*

the fine job done by Ben and his team and subsequently by John and Dan Dilley.

The Sabre did not remain in the ownership of John Dilley for very long, being acquired in 1990 by the Golden Apple Trust. Fort Wayne Air Services (FWAS) of Indiana were commissioned by the Trust to restore the cockpit to its original condition (which included the acquisition of a working ejection seat) before being transferred to the UK at the end of 1991.

BELOW Ben Hall gives Jim Larsen a ride in 48-178 following the restoration. *(Jim Larsen)*

ABOVE AND RIGHT When Ben Hall first restored No. 178 it lacked an ejection seat. Instead, an arrangement was fashioned whereby a passenger could be flown. The 'lucky' person along for the ride would strap into a bench-type seat whereupon the pilot would then fold down a second seat and climb in on top. Other than the official two-seat TF-86F trainer (of which only two were built), 48-178 was the only other Sabre to fly with two people on board. Later when in the hands of the Golden Apple Trust the cockpit was overhauled and the ejection seat reinstated.

48-178 operation in the UK

The Golden Apple Trust was founded with the vision to restore, preserve and demonstrate icons from the machine age, a period of rapid development in a time before computers would revolutionise vehicle and aircraft design and performance. This objective was even more important given that, at the time, little effort was being made to preserve such vehicles, let alone in working order. Historic competition racing cars and piston-engined aircraft from the machine-age era were already catered for by wealthy enthusiasts and military-funded spin-offs such as the Battle of Britain Memorial Flight. The public had reasonable access to these cars and aircraft at race weekends and air shows. However, apart from a Gloucester Meteor, one or two de Havilland Vampires and some Hawker Hunters, very few historic jets were publicly displayed.

The Golden Apple Trust was fortunate to have as trustees the motor racer, enthusiast and businessman Martin Colvill (co-owner/founder of the world's largest Lotus distributor) and Duncan

Simpson OBE, former chief military test pilot at Hawker Siddeley. Coincidently, early in his test-pilot career Duncan had evaluated the Sabre for the RAF, which went on to buy 431 pending the arrival into service of the Hunter.

Research was undertaken by the Trust to establish what was missing in the UK and at this time the opportunity of acquiring an airworthy F-86A Sabre, the first of the swept-wing jets,

ABOVE 48-178 as she appears today. Now registered in the UK as G-SABR, this photograph was taken as she is taxiing out for take-off (followed by a de Havilland Vampire) at the air show at RAF Leuchars (St Andrews) in 2010. *(Mark Wright)*

LEFT In 1989 The Golden Apple Trust were presented with the opportunity of acquiring the only remaining airworthy F-86A, the first of the swept-wing jets. *(Author)*

F-86A-7 G-SABR
Golden Apple Operations, Duxford
2011

arose which perfectly matched the aims of the Trust. The aircraft had recently been bought by John Dilley of FWAS, Indiana, from Ben Hall. With manual reversion on all the controls an 'A' model (from an operational point of view) was acceptable to the Civil Aviation Authority (CAA) and historically it was important as the only example of the variant. Also the 'A' model with its higher aspect ratio wing was considered more aesthetic than the later 6-3 variants.

Following inspections by Chief Engineer Eric Hayward of Jet Heritage and a test flight by Chuck Scott (Ex-USN F-8 pilot), 48-178 was transferred to Golden Apple around the turn of 1989–90. As FWAS had come to know 48-178 well and had good engineering expertise, it was agreed that they would carry out the work

required to revert the aircraft to a stock cockpit configuration with radar ranging gun sight and a radio suite cleverly concealed in the right-hand console. The USAF Museum at Wright-Patterson AFB provided considerable assistance in the task of restoring the cockpit. A new Mitsubishi-built ejection seat was fitted, which was a more up-to-date example, but kept the authentic looks demanded by the restoration. Work was also done on the engine and hydraulics.

When the time came to move the Sabre to the UK there was a host of pilots volunteering to fly her across the Atlantic, many with no Sabre time whatsoever! It was decided to take the sensible option and ship her across to Southampton in two 40ft containers, one containing the aircraft and wings and the other with the spare engine and numerous other parts.

On 6 January 1992 the containers arrived at the Jet Heritage hangar in Bournemouth and the aircraft was carefully reassembled. By late spring she was ready for engine runs and Adrian Gjertsen flew her first flight from a UK runway on 21 May 1992.

A Sabre had not been seen over the UK for many years so No. 178 was in immediate demand from air show organisers, with Biggin Hill being the debut. Sabres are special for Biggin Hill as Jock Maitland (chairman of Air Displays International) had flown the type in the Korean War as an exchange pilot and had claimed a victory over a MiG-15. Jock was kind enough to write a piece in the programme describing the air war, but lamented that just as his parents proudly had a picture of him over

their fireplace, there would be somewhere a similar photograph of the young Russian pilot who did not return.

Golden Apple had always felt that ultimately the Sabre should be based at the Imperial War Museum, Duxford, where about ½ million visitors a year would have access to it. Mark Hanna, then chief pilot of the Old Flying Machine Company (OFMC) based at Duxford, had quickly recognised No. 178 as one of the most glamorous aircraft on the display circuit and pitched hard for the Trust to move her to Duxford under the OFMC umbrella. After careful evaluation of the facilities, runway and engineering skills available, the decision was made to move the Sabre from Bournemouth to Duxford where OFMC flew her for some years, often in the company of their two-seat MiG-15, making a very evocative duo.

Following Mark's very sad death in Spain at much too young an age, at the controls of OFMC's Me109 Buchon, the Sabre moved to the care of the Aircraft Restoration Company (ARC) also at Duxford, and a new chapter began for No. 178.

Golden Apple was delighted to have Air Marshal 'Cliff' Spink as its new chief pilot and a new arrangement was made with ARC as a partnership in which they are responsible for all engineering activity and Golden Apple Operations, with its own CAP632 authorisation, for the flying side.

Cliff was soon able to gather around himself a small team of pilots, namely the author (an ex-RAF Harrier pilot who had flown No. 178 and the MiG-15 for several years when they were operated by OFMC), Keith Dennison (then engaged in testing the Typhoon for BAE Systems) and Dave Harvey (one of the RAF's most experienced Hawk pilots). After a few seasons the author took over from Cliff as Golden Apple's chief pilot.

The engineering side at ARC was equally robust, with experienced engineer Colin Swann leading the team. Personal responsibility for the aircraft was assumed by George Francis and then by Joe Kennedy. More recently Andrew Foster has taken on responsibility for the maintenance of the aircraft. Many others who work in the office, the hangar and on the airfield have worked hard and shown a 'can-do' spirit to overcome many, many complex challenges so that the aircraft remains serviceable.

ABOVE Cliff Spink flies G-SABR on a post-maintenance test flight. Note that the under-wing drop tanks have not yet been fitted. *(Colin Norwood)*

LEFT The author together with Joe Kennedy (formerly of the Aircraft Restoration Company) and fellow pilot Cliff Spink at the Golden Apple training day in 2009. ARC provides all the engineering support and services needed to operate the Sabre.

Chapter Four

Flying the F-86A

The Sabre can be justly described as an aircraft that was way ahead of its time. Whenever 48-178 is taken into the air it is hoped that the viewing public will be reminded of just how ground-breaking and revolutionary this aircraft was. It is indeed fortunate that the Sabre can be demonstrated in flight, the only environment that really shows her true capabilities.

OPPOSITE 48-178 shows off her lines. *(John Dibbs)*

entering cockpit

Golden Apple Operations' F-86A (48-178 (G-SABR)) is a typical 'A' model, being somewhat unrefined and underpowered compared to later variants. The early Sabres had many adverse handling qualities, which were mostly solved by the introduction of the F-86E. The fact that the 'E' flew in September 1950, a little over two years after the first production 'A' model entered service, illustrates the rapidity of development at that time.

Later J47-powered variants of the F-86, together with licence-built versions fitted with Avon and Orenda engines, enjoyed much more power in addition to better flying qualities. If there is an advantage over later versions it is that the 'A' model was able to revert to manual flying control thereby making it the only Sabre variant that could be flown without hydraulics: in the eyes of the UK CAA (at the time 48-178 was imported to the UK) this fact made the Sabre a viable proposition for private operation.

The allure of flying 48-178 comes from the very idiosyncrasies that were developed-out in later models. The 'A' model certainly had its faults, but it is still an amazing aircraft, even by today's standards. These particular quirks and foibles remind us that this undisputed icon of the jet-age must never be taken for granted.

This account is intended only to give a flavour of what it is like to fly 48-178. Although it can't possibly convey the entire experience it is hoped that it will provide an insight and be informative. I am (obviously) not a qualified test pilot and those readers who are might consider that some of the aerodynamic explanations are being oversimplified. The basic descriptions given on this subject, as they relate to the Sabre, are purely a reflection of my limited understanding of the complex stability and momentum interactions at play.

What can be described as 'corporate knowledge' on the Sabre is very narrow. It is not possible to enjoy the obvious benefit derived from a crew room full of Sabre pilots, all current on the type, all sharing their experiences. Crew-room 'banter' and shared first-hand accounts have always been an effective way for fledgling fighter pilots to learn their trade. It's just as well that the aircraft is flown well within its flight

The UK CAA allows the flying of ex-military aircraft on the civilian register under the umbrella of CAP 632 (Operation of 'Permit-to-fly' Ex-military Aircraft on the UK Register). This regulation classifies aircraft in the following categories:

Simple – single-piston-engine types.

Intermediate – multiple-piston-engine or turbine- (single or multiple) engine types with simple mechanical flying controls or with power controls having an independent back-up system which ensures continued safe flight.

Complex – all other types, in particular those types having features that require a high degree of specialised knowledge and equipment to maintain (for example, types with no independent back-up system to powered flying controls or with auto-stabilisation systems or electronic engine controls).

Complex types; for example, the Avro Vulcan, Blackburn Buccaneer or English Electric Lightning, would all require (expensive) specialist industry support and are hence not practical for private ownership.

The F-86A is classified as being an Intermediate type. Since 48-178 was imported in 1992 the regulations have been modified so that now later variants of the Sabre could (in theory) also be operated in the Intermediate category, although none are. The classification is the responsibility of the CAA.

envelope, in good weather and in a benign environment with no one shooting at it.

Our flight in the Golden Apple Trust's unique F-86A begins with the exterior inspection.

Pre-start checks

The walk-around inspection starts at the left wing root and goes clockwise (viewed from above) around the aircraft: in principle it is no different to the kind of external inspection on any other aircraft. The point is to check for obvious signs of damage, unsecured hatches, loose panels, leaks or for anything else that might be amiss. This will have already been done by the engineer who preflighted the aircraft, but in aviation there has always been a culture of double-checking. It's a case of 'never assume, check': this approach has nothing to do with there being a lack of trust; it's just a sensible ritual.

The underneath of the aircraft is always examined closely for any evidence of leaks. There is never more than the tiniest amount of hydraulic fluid present and then typically only if the aircraft has been standing cold for some period of time. Strictly speaking this is not a leak, but more of

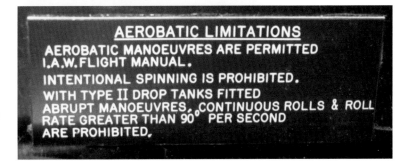

WARNING

THIS AIRCRAFT DISPLAYS MANY OF THE DEFICIENCES IN STABILITY OF THE ORIGIN--AL TYPE DESIGN.

ALL MANOEUVRES LIKELY TO LEAD TO AN INADVERTANT DEPARTURE FROM CONTROLLED FLIGHT ARE PROHIBITED.

AEROBATIC LIMITATIONS
AEROBATIC MANOEUVRES ARE PERMITTED I.A.W. FLIGHT MANUAL.
INTENTIONAL SPINNING IS PROHIBITED.
WITH TYPE II DROP TANKS FITTED ABRUPT MANOEUVRES, CONTINUOUS ROLLS & ROLL RATE GREATER THAN 90° PER SECOND ARE PROHIBITED.

LEFT AND BELOW
Placards directly in cockpit serve as a reminder (if one were needed) that the F-86A is 'special' and has certain limitations that must be observed.
(Author)

a trait: once the system pressurises then all the various seals tighten up and any seepage stops. It's very important to check the air intake for any objects that might damage the engine on start.

EXTERIOR INSPECTION

① **NOSE.** Starting at nose of airplane, make following checks:

  a. Nose gear ground safety lock removed. Tow pin safety cap tightened.

  b. Nose gear oleo strut extension; tire for slippage and proper inflation. Check for hydraulic fluid leaks.

  c. Intake duct clear – pitot tube uncovered.

  d. Armament doors secured; gun port plugs installed.

② **RIGHT WING.**

  a. Slats for freedom of movement.

  b. Fuel caps, oil and hydraulic tank access doors secured.

  c. Drop tank clearance; sway braces for looseness.

  d. Wing tip and navigation light. Pitot head cover removed (F-86A-6 and F-86A-7 Airplanes).

  e. Aileron for freedom of movement, loose rivets, etc; flaps for position.

  f. Gear oleo strut extension; tire for slippage and proper inflation; wheel chocked.

  g. Any leaks of fuel, oil, or hydraulic fluid. Check hydraulic accumulator pressure (right wheel well).

  h. Gear door position.

③ **RIGHT FUSELAGE.**

  a. All access doors secured.

  b. Speed brake in; condition of skin around brake. Check for hydraulic leaks.

  c. Duct in oil access door for evidence of dust streak (indicating a faulty generator).

④ **EMPENNAGE.**

  a. Tail surfaces for condition; elevator for freedom of movement.

  b. Navigation lights.

  c. Rudder trim tab neutral.

  d. Tail-pipe cover removed; tail pipe for cracks or excessive distortion.

  e. Check aft fuselage aspirator fuel drain for freedom of movement.

⑤ **LEFT FUSELAGE.**

  a. Make same checks as on right side, except step c.

  b. Destructor plugs in IFF equipment.

⑥ **LEFT WING.**

  Make same checks as on right wing, only in reverse order, and, in addition, check aileron trim tab in neutral.

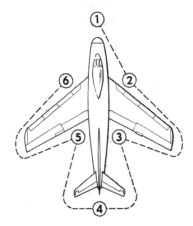

Inspection of the tyres and brake-wear indicators requires crawling under the aircraft and, while on hands and knees, the operation of the fuel pumps can be verified (via a pair of test switches in the left-hand wheel bay) together with the hydraulic system accumulator pressure gauges (located in the right-hand wheel bay). The nitrogen bottle pressure gauge in the nose-wheel bay is also checked: the nose leg unfolds forward into the airflow and this compressed gas is the source of power to 'assist' it being lowered in the event of a hydraulic failure; for the emergency extension of the main wheels gravity is enough. Particular attention is paid to the correct fitting of the fuel caps (there are six), especially those that provide an airtight seal for the drop tanks. Failure of either drop tank to pressurise will prevent fuel from transferring as well as create an imbalance. One thing that can be done on this aircraft (unlike any other Sabre variants with irreversible hydraulic powered controls) is to grab hold of the flying control surfaces (elevator and ailerons – it isn't possible to reach the rudder) and check the movement for smooth and correct operation.

The Sabre offers a commanding seating position with entry into the cockpit being the first of many challenges facing the pilot: it's a long way up. In the absence of external steps the technique involves a climbing and balancing act accomplished with the aid of a 'kick-in' step

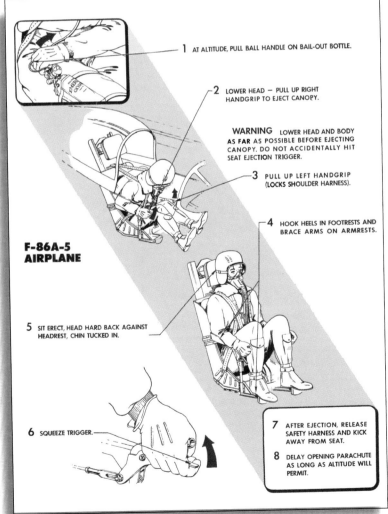

F-86A-5 AIRPLANE

1 AT ALTITUDE, PULL BALL HANDLE ON BAIL-OUT BOTTLE.

2 LOWER HEAD – PULL UP RIGHT HANDGRIP TO EJECT CANOPY.

WARNING LOWER HEAD AND BODY AS FAR AS POSSIBLE BEFORE EJECTING CANOPY. DO NOT ACCIDENTALLY HIT SEAT EJECTION TRIGGER.

3 PULL UP LEFT HANDGRIP (LOCKS SHOULDER HARNESS).

4 HOOK HEELS IN FOOTRESTS AND BRACE ARMS ON ARMRESTS.

5 SIT ERECT, HEAD HARD BACK AGAINST HEADREST, CHIN TUCKED IN.

6 SQUEEZE TRIGGER.

7 AFTER EJECTION, RELEASE SAFETY HARNESS AND KICK AWAY FROM SEAT.

8 DELAY OPENING PARACHUTE AS LONG AS ALTITUDE WILL PERMIT.

LEFT The ejection seat in 48-178 was manufactured in Japan by Mitsubishi and taken from an F-86F built under licence by the same company. The seat is still basic by modern standards, but it is fully functional and has the correct look and armour protection of the original. When compared to modern-day zero-zero rocket seats the Sabre's has poor performance and both height and speed are required for a successful escape. It also lacks the simple operation and automation of later-generation seats, thus requiring some premeditation (such as lifting feet into stirrups), plus various bodily contortions before it can be used. It is also very important to duck down before firing the canopy remover. On the upside the sitting position is spacious and actually quite comfortable thanks to adjustable seat height and variable rudder pedal extension.

panel and a retractable grab handle positioned perfectly to do serious damage to delicate parts of the body in the event of a missed footing. Dismounting is equally hazardous: it's a long way down. The destination is well worth the precarious journey as the Sabre's cockpit is as good a place to be as any. Once standing on the seat facing rearward the ejection seat can be inspected then the canopy jettison and seat firing pins are removed: it's impossible to do this once strapped in and you only forget this fact once. The two other safety pins protect the canopy jettison handles and the seat firing triggers: these are easily reached and are removed once strapped in.

Thanks to the low sill and the 'bubble' canopy, the visibility up, down and to the rear is exceptional, indeed it's better than many modern day-fighters. The slightly convex rear-view mirror mounted directly in front of the pilot only enhances this by eliminating any 'blind spots' to give a true 360 degrees of lookout. The Sabre pilots' view completely outclassed that of their MiG-15 counterparts. The only area where visibility is not brilliant (especially in a rain shower, where it is more or less non-existent) is directly out the front, this being primarily because of the thick and heavy 'vee'-shaped windshield, bulletproof glass and broad canopy arch.

Given that the science of ergonomics was yet to be invented, NAA did a very good job indeed with the cockpit layout. Controls and instruments are well placed, logical and (with only one or two exceptions) easy to use and interpret. By comparison the MiG-15 was totally inferior in this respect. The MiG was chaotic, confusing and agricultural. If the Sabre were to be likened to a thoroughbred sports car, where everything just feels right, then the MiG-15 would have to be equated to a farm tractor. British fighters of the period were not much better laid out than the MiG, which makes the thoughtfulness of the NAA designers even more impressive.

There are almost no electronics in the F-86A and nothing even remotely resembling a computer. Everything is truly pilot controlled, this being especially true of the engine. This is highly susceptible to surging across the whole operating range and is very slow to accelerate: it's also easy to 'cook', especially during start. In some respects engine handling and the starting procedure is one that requires the most concentration.

Engine starting sequence

For engine starting an external power source is definitely preferred: starting on batteries is marginal and all the help available is required. Under any circumstances starting with a tail-wind is to be avoided, such is the sensitivity of the starting sequence to hot starts.

Once all the usual pre-start checks have been completed the technique for getting everything turning is as follows:
- Press the starter button: this engages the starter motor.
- As soon as engine motoring is observed (about 4% rpm) very slowly open the throttle/HP cock until fuel pressure is observed (about 10psi is required). This is a two-handed job as the HP cock is incredibly

WARNING

- Any one start or acceleration during which exhaust temperature exceeds 1000°C momentarily, or any 10 starts or accelerations during which exhaust temperature exceeds 870°C, shall constitute overtemperature operation and require that the engine be removed from the airplane and returned to overhaul (required in case of 10 hot starts, regardless of time lapse between starts). The temperature and duration of all overtemperature operation (870°C) shall be entered in Form 1. Five starts or accelerations during which exhaust temperature exceeds 870°C requires that the engine be carefully inspected for possible damage prior to flight.
- When a hot start occurs, shut down engine immediately. If smoking or fire persists, engage the starter for approximately 20 to 30 seconds to rid engine of excess fuel.

sensitive and it is very easy to over fuel at this time (which could result in one destroyed J47 and a very large bill for a replacement).

■ When light-up occurs (indicated by a rapid rise in jet-pipe temperature, although it can be heard too) it's a good idea not to be looking down the jet-pipe: a lick of flame extending several feet out of the jet-pipe is not uncommon.

■ Patience is now crucial as (very) gradually more fuel is trickled in and modulated to keep the jet-pipe temperature within limits (690°C) as the rpm creeps slowly up to idle. It needs to be nursed along because the engine is struggling throughout the process

not to surge. All kinds of strange groaning noises and resonations together with odd vibrations add to the drama.

■ Eventually the starter cuts out and the motor changes mode to a generator.

■ Finally the engine (in its own time) comes up to idle.

During start the engine is under a great deal of stress with no automatic systems to protect it from damage. The consequences of rushing or getting the sequence wrong can be more than simply embarrassing. There are a number of potential starting problems that it is necessary to be aware of:

■ 'Hung start': engine lights up but fails to accelerate. These are common and often experienced during an internal battery start.

■ 'Hot start': engine reaches turbine temperature limits with a low rpm. In some respects every start is a 'hot start', albeit one carefully controlled by the pilot as he coaxes the beast into life.

■ 'Wet start': fuel goes in but no ignition takes place. This is less common and results in a puddle of fuel swamping the jet-pipe. Before attempting another start this excess fuel must be blown through and the remainder mopped out. The aircraft sits slightly nose down so it's not possible to completely drain all of the fuel and the start that follows is usually a good photographic opportunity thanks to the torrent of flame that temporarily turns the jet-pipe into a blow torch.

CAUTION

IF IGNITION DOESN'T OCCUR WITHIN 10 SEC-
ONDS OR WHEN ENGINE RPM REACHES 9%,
CLOSE THROTTLE AND PUSH STOP-STARTER
BUTTON. ALLOW FUEL TO DRAIN 3 MINUTES
BEFORE ATTEMPTING RESTART.

Once the engine is running, the hydraulic system can be set up, which in many respects is the weakest on the aircraft. The hydraulic system is pressurised by a *single* pump delivering 3,000psi to a *single* line via a *single* reservoir, all without any notion of a priority system to protect vital flying control operation or lacking an effective accumulator to take up any slack. This all means that it is necessary actively to manage this particular resource because, even though the aircraft is quite flyable without hydraulics, it is not desirable. It's fair to say that aeronautical designers would never propose such a solution nowadays.

With the engine now driving the hydraulic pump the protracted process designed to bleed air from the system is undertaken. This involves holding the flying controls at their maximum deflection and bypassing the aileron boost while the hydraulic pump purges any air within. Like the brakes on a car, an airlock in the system will cause unresponsiveness (or sponginess) at best, or complete failure at worst. Neither scenario is conducive to enjoying the pleasures of flying a high-performance jet-fighter.

Speed-brake operation is checked (there is no indication in the cockpit of speed-brake position so a visual check from the ground crew is required), the landing gear supplementary doors are retracted (by closing the 'Doors' switch) and the elevator and aileron boost pressures are checked together with a test of the aileron boost bypass system. The management of the hydraulics thereafter, and especially in flight, is principally always to ensure that adequate pressure exists to protect the flight controls (as well as on the ground the nose-wheel steering). The system simply cannot provide enough pressure to satisfy multiple simultaneous demands.

The rest of the pre-flight preparations are straightforward and consist of switching on radios, checking various sub-systems, erecting instruments and configuring the aircraft for take-off. At some point before taxiing four fingers are held up to the ground crew to confirm that all four of the ejection seat safety pins have been removed and stowed.

The final action is to ensure that the canopy is closed. When doing this it's a good idea to duck forward to avoid being knocked on the head by the canopy arch.

Taxying

As already inferred the Sabre has some endearing and a few not so pleasing idiosyncrasies: the waywardness of the nose-wheel steering is one example of the latter. The system is engaged by pressing and holding down a button on the control column that connects the rudder pedals to the nose-wheel leg pivot allowing steering within a limited range. At higher taxi speeds directional control is possible using differential braking alone; however, when manoeuvring at a slower pace the nose-wheel steering is the key to precise control with minimum power (always a consideration with jet aircraft). For the system to 'hook-up' the nose-wheel must be more or less pointing straight ahead. If outside of a very

limited range then engagement is impossible and the aircraft is effectively marooned on the taxi-way, unless there is plenty of room and a handful of power can be applied.

Another poor design feature is the nose-wheel steering engagement button (located on the control column). This must be held 'in' to activate the system as releasing this spring-loaded button disconnects the interlock with the rudder pedals. The inconvenience here is that the pressure required to keep this button engaged is quite high and consequently the finger chosen for the task will at first become very numb, then distractingly painful. If you want to experience this sensation for yourself then place a tic-tac mint (or similar small, hard and round object) between the tip of your index finger and your thumb, now squeeze tightly together and hold for 5 minutes.

Take-off

Lined up on the runway there are just a few more important things to do before rushing off in a cacophony of earth-shattering noise and climate-changing smoke. The first is to check the engine fuel-control systems. The primary system is very basic; the back-up is even more primitive being intended only to provide enough control to 'get you home'. Before take-off both these systems are tested. Engine acceleration is woefully slow from idle and at all times has to be carefully controlled to avoid a compressor surge or over-temperature. Whenever the throttle is advanced attention is split between the exhaust gas temperature gauge (to ensure the limits are not exceeded) and the rpm indicator to check that it is increasing. The engine is kind enough to give out an aural warning by way of a friendly rasping noise in advance of a not so gracious full-blown compressor stall. This is just as well as there are no governors or other gizmos to prevent such an occurrence. In other words, the health of the engine is controlled by the pilot's left hand and nothing else. Once above about 70% rpm the acceleration and handling become much more acceptable although never 'care-free', as in modern aircraft.

Finally, the potentially life-changing check

of hydraulics and trim is completed and with brakes released the take-off roll commences. Acceleration cannot be described as rapid, but it is smooth and steady. Cross-winds present no real problem and with the rudder becoming fully effective at 50 knots the nose-wheel steering button can be released, thereafter controlling direction aerodynamically with the rudder.

At about 100 knots the technique is to pull back on the stick enough to take the weight off the nose-wheel. The stick forces are quite high and any temptation to pull too hard must be resisted. This aircraft must not be rotated too nose high as that only adds a bucketful of drag to the competing forces struggling to get airborne. The action here is more analogous to taking the weight off the nose-wheel rather than rotating to a flying attitude. The Sabre sits a little nose down on the ground so the primary purpose of lifting the nose a few degrees is to introduce the wing to some lift-giving angle of attack. At 110–15 knots the aircraft lifts off smoothly, albeit in a fairly nose-low attitude.

With a positive rate of climb the landing gear is selected up and at about 150 knots the flaps are retracted. Once in the air the Sabre picks up speed fairly quickly to the extent that the nose needs to be lifted more steeply to keep within the gear and flap limiting speed of 186 knots. Throughout the take-off run at least one eye must be kept on the engine gauges to correct any rpm creep or turbine over-temperature, either of which will require manual intervention to keep within limits.

General handling

It is common for air-show commentators to chant the 'if it looks right, it flies right' mantra when describing the F-86. They are (of course) partly right, but also surprisingly yet understandably wrong. There is no doubting that she 'looks right': no question about that. However, 48-178 is an 'A' model and as such is nowhere near as well refined as later models in the series. When the British naval test pilot Capt Eric 'Winkle' Brown RN described the Sabre as 'the finest jet aircraft I have ever flown from a handling point of view', he was referring to the F-86E. In fairness, 'Winkle' Brown was impressed by the 'A' too, but would

ABOVE The J47 is a very smoky and somewhat unrefined first-generation axial-flow turbo-jet engine. During the Korean War the allied forces' primary friend or foe recognition feature was that 'if it smoked it was friendly': the MiG-15, although similar in profile, used a copy of the Rolls-Royce Nene, which was a clean-burning centrifugal flow engine. Smoke adds to the Sabre's uniqueness on the air-show circuit. *(Huw Hopkins)*

LEFT 'If it smokes, it's friendly' – the Sabre displaying at Lelystad, Netherlands, in 2007. *(Tim de Groot)*

in all likelihood not have made such a definitive statement about the earlier version, such were the improvements brought about by the 'E'. It's not that the aircraft doesn't fly well it's just that there are many aspects to its handling that would be deemed unacceptable nowadays. These traits existed principally because lessons were still to be learned about swept-wing and high-speed aerodynamics and so none of these deficiencies should be surprising given the Sabre's place in history. There is no criticism intended of the era's pioneers of high-speed flight either: it just needs to be said that there is a common misperception today, by some, that the F-86A was completely 'sorted', whereas it was not.

For its time the Sabre's performance must have been a revelation. Acceleration and climb rate in particular were in a totally different class to anything that had gone before. The swept-wings provided not only less drag, but also much better manoeuvrability at high altitude/high Mach number than a straight-wing aircraft. However, when compared to later Sabres, and other third-generation jet-fighters such as the Hunter, there were many handling shortcomings.

48-178 is an aircraft that changes its character quite distinctly depending on how fast (and how high) it's flying. The 'A' model Sabre is actually not that nice to fly at slow speeds or at high Mach number/high altitude. One issue is the unsubtleness of the hydraulic boost system that can sometimes make the pilot feel a little disconnected from the aircraft and, for dynamic stability reasons, there are a few corners of the envelope that it is best to stay away from. That said she is a delight when flown smoothly in the range of 200–400 knots in the thick air at lower levels.

Without delving too deeply into the laws of aerodynamics, there are a whole host of factors that come into play when flying slowly: unfortunately these coincide at around about landing speed and conspire to make life difficult for the pilot. Poor engine response, high angle of attack, increasing drag (it is very easy in this aircraft to get 'behind' the drag curve, quite literally), weak pitch stability and limited visibility out the front all require the utmost attention from the pilot. Get a little bit too slow or low on the landing approach and you get that swept-wing sinking feeling, which if not checked will

ABOVE The author leads Jonathon Whaley in his ex-Swiss Air Force (J-4104) Mk 58A Hunter 'Miss Demeanour' (G-PSST) for a flypast at the Duxford Autumn Air Show in 2009. Hunters were first delivered to Switzerland in 1958, with this particular Hunter being from the second contract in 1971. The rapid pace of development during the 1950s is very well illustrated by the fact that the Hunter (although it had teething problems) was coming into service less than seven years after the first flight of the XP-86, yet it was a much better aircraft by almost every measure. *(Gary Brown)*

RIGHT This chart shows the elevator stick forces required to maintain a level attitude as Mach number increases with trim set for Mach .6. The aircraft is stable up to about Mach .8 with a gentle push force required. At Mach .85 a mild force reversal occurs as the aircraft tends to tuck nose down. At Mach .9 push forces are again required and the aircraft tends to tuck nose up. This is the result of shock waves and airflow separation occurring on the upper wing surface.

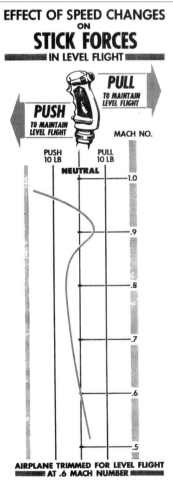

EFFECT OF SPEED CHANGES
ON
STICK FORCES
IN LEVEL FLIGHT

PULL
TO MAINTAIN LEVEL FLIGHT

PUSH
TO MAINTAIN LEVEL FLIGHT

MACH NO.

PUSH 10 LB   PULL 10 LB

NEUTRAL
1.0
.9
.8
.7
.6
.5

AIRPLANE TRIMMED FOR LEVEL FLIGHT
AT .6 MACH NUMBER

run you out of power and back-stick in a heart-beat, with the inevitable consequences. It is not helped that G-SABR has had her slats locked in, which was one aerodynamic feature that partly solved some of these problems. However, the poor slow-speed handling characteristics should not be overstated because those who fly the aircraft are all highly experienced ex-military fast-jet pilots who understand the problems and know exactly how to keep themselves out of trouble and thus ahead of the drag curve.

A good thing is the Sabre's benign stall characteristics. The aircraft gives plenty of warning in the form of a gentle buffeting on the tail plane (felt through the control column) a few knots before becoming fully developed and even then remains controllable throughout. There is no 'bite' or wing drop and, although the rate of the descent drops off the clock, recovery is standard and straightforward. The slightest check forward (to reduce the angle of attack) is all that is needed to recover from the stall. However, in a stall a high rate of descent quickly builds up and therefore stalling at low altitude, unintentional or otherwise, is to be avoided. The excellent warning of an approaching stall is just as well because on final approach, where (by definition) there is not much height above ground, combined with a high sink rate (which rules out any successful ejection option) plus very poor engine responsiveness (which lessens the chances of a successful go-around), a stall in this territory is a once only, never to be repeated, exercise.

Above about 200 knots it's a different story: the aircraft is delightful. She is highly responsive and generally well balanced.

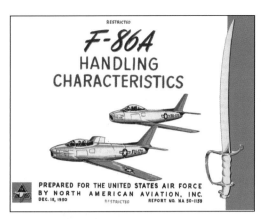

In stability terms the aircraft is dynamically stable most of the time (unstable in pitch under certain circumstances where there is a manageable tendency to 'dig in' to a turn, a trait that incidentally is evident in the MiG-15 to a seriously dangerous degree), statically stable (although marginally so in pitch), but that said mostly behaves the way one would expect.

G-SABR is no longer operated in the high-altitude transonic speed range. However, by all accounts in this regime the 'A' model was twitchy in roll and generally poor in pitch due to an ineffective elevator requiring large control inputs. Furthermore, at high speed the speed-brakes are known to give quite a pitch-up trim change when extended.

The horizontal tail surfaces on the 'A' are of conventional design and (on all but very early Sabres) there is an artificial feel system

LEFT AND BELOW
The handling characteristics of the F-86 were so unlike any aircraft that had gone before that in December 1950 a book was produced by NAA for the USAF intended to highlight these differences. The foreword to this book acknowledges that, 'Numerous accidents with high-speed jet airplanes have been attributed directly to pilot misunderstanding or unfamiliarity with specific handling characteristics and limitations.'

Warning

OPENING SPEED BRAKES AT HIGH SPEEDS IMPOSES AN ADDITIONAL 2G. THEREFORE, IF SPEED BRAKES ARE OPENING, MAKE ALLOWANCE WHEN PULLING STICK BACK OR USING NOSE-UP TRIM, TO AVOID EXCEED-ING G-LIMIT.

WARNING

Except in extreme emergencies, do not pull the stick back or use nose-up stabilizer trim during the time the speed brakes are extending. To do so may result in exceeding the limit load factor. The speed brakes open fully in approximately 2 seconds at high rpm. (Refer to paragraph 6-37.)

which comprises a bob weight incorporated into the elevator control system. This bob weight adds about 6lb of stick force per 'g' and was incorporated to prevent the pilot from overstressing the airframe. Like many swept-wing aircraft the Sabre has a tendency to 'g' overshoot (sometimes known as 'dig in') when manoeuvring in the buffet. This characteristic is caused by the basic instability of the aircraft near to its maximum angle of attack. As 'g' increases the tip of the wing does not increase its lift, but instead enters a partially stalled condition. Relatively more lift is then provided by the inboard part of the wing and, since it is swept, the centre of lift moves forward causing

the nose to pitch up even more, or to 'dig in'. In addition, the flow of air off the inboard portion of the wing is deflected down more and this flow passes over the tail, which adds to the nose-up pitching tendency.

The handling of the Sabre was much improved by the introduction of a more powerful and responsive all-flying tail plane on later variants, but in the 'A' model it is enough to say that when flying inside the self-reduced envelope of operation today (where high transonic Mach numbers are not a factor) she is viceless and straightforward to fly.

The aircraft is quite manoeuvrable, especially in pitch (even without the slats), although as the

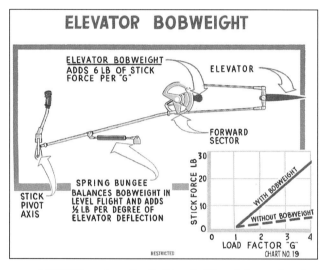

ABOVE To reduce the likelihood of the pilot overstressing the airframe and to improve the feel in pitch during hard manoeuvres a bob weight was incorporated: this increased the stick forces as 'g' increased.

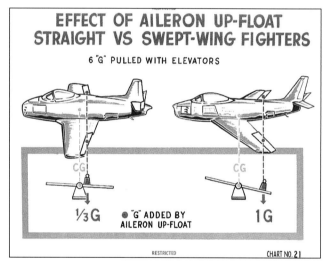

ABOVE Unique to the 'A' model is yet another characteristic that contributes to 'g' overshoot: this is aileron upfloat and it is present at high Mach number and high 'g' due to air loads on the lower surface of the wing pushing the ailerons up. Although tie-across cables prevent the ailerons from moving in the same direction, at high loading these cables stretch slightly allowing a few degrees of up-float. The ailerons are more than twice the size of the elevators and half as far aft of the centre of gravity, therefore approximately one degree of aileron up-float causes about the same pitch up as one degree of elevator. Sabres with irreversible controls were not affected by aileron up-float.

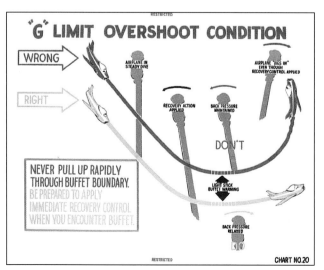

LEFT Even with the bob weight installed it was still important for the pilot to avoid over-pitching. For example, during the recovery from a dive the correct technique is to relax back pressure on the stick as soon as light buffet is encountered, not the most instinctive of responses for a pilot diving toward the ground.

speed reduces so too does the pitch stability. Hard pitching manoeuvres at slower speeds, such as pulling over the top of a loop, need to be flown carefully to avoid 'dig in' as previously described. While nowhere near as bad as some early swept-wing fighters (notably the MiG-15) this trait is still evident and needs to be respected.

A specific characteristic of the Sabre was the relatively large size of the ailerons (almost half the span, and with a broad chord). The aircraft's roll rate is certainly very good and fortunately there is very little adverse yaw provided the aircraft is 'unloaded' during the rolling manoeuvre. (In some aircraft types drag from the ailerons during a rolling manoeuvre can cause a dangerous yaw to develop. This is made much worse if the wing is producing large amounts of lift as the drag on the up-going wing will then be much higher to the point that the yawing moment overcomes the aircraft's directional stability.)

The pilot's flight manual ominously states that above 250 knots roll control is 'impossible' following a failure of the aileron boost. This is because the air-loads on the control surfaces at high speed would prevent any worthwhile movement. That said, a measure of roll control using rudder is still possible and reasonably effective as with many swept-wing types. (In a swept-wing aircraft any yaw tends to increase the lift on the wing that is now less swept relative to the airflow. This then tends to roll the aircraft in the direction of the yaw. To roll to the right, yaw the nose to the right.) Later variants of the Sabre with dual hydraulic systems

and irreversible controls solved the problem of dependency on a single-system failure; however, the downside was that a double hydraulic failure rendered the aircraft completely unflyable due to a lack of any manual reversion.

The aileron boost is another quirky system that testifies to the 'A' model Sabre's first-generation swept-wing design status. Movement of the control column against the air loads creates a resistance that the pilot pushes against. Just as the pilot begins to wonder if the system is actually working and pushes a bit harder the boost kicks in and delivers a helping hand to the tune of 37 times the input of the pilot (50 times on very early aircraft). In other words, for every 1lb of force applied to the stick the boost system applies 37lb of force, and all in the blink of an eye: there is nothing progressive or subtle about this system. Precise and smooth roll control, especially at high speed, is quite difficult as a consequence and can lead to over-control until experienced. By comparison, the elevator is much less boosted at seven times pilot input.

I often think that in 1949, without the benefit of a two-seat trainer or an extensive pool of experience, early F-86 pilots transferring from types such as the P-51 or even the P-80 with their straight wings and better power response must have experienced at least one or two unpleasant surprises during their conversions. The fact that in the early days accidents involving Sabre landing 'short' were quite common is evidence of just how steep the learning curve must have been.

The casualty rate during training was so high that many student pilots were often teased by their instructors that, 'If you ever see the flag at full staff, take a picture.'

One thing you will not see an F-86A pilot doing very much of is negative 'g' manoeuvres or prolonged inverted flight. The secondary effect of being upside down (for any more than a few seconds) is almost always a temporary hydraulic failure. This happens due to fluid that normally covers the outflow pipe in the bottom of the hydraulic reservoir being replaced with air as the fluid swaps ends. This air is then sucked through the system and ends up at the hydraulic pump. Air, unlike incompressible hydraulic fluid, is compressed very nicely indeed to the extent that the hydraulic pressure drops to almost zero. Thanks to the system's self-bleeding nature the pressure eventually comes back, but not before the roll characteristics using rudder alone have been explored, a plan to land with limited brakes or no nose-wheel steering considered and the procedure for lowering of the landing gear by gravity has been reviewed from the flight reference cards.

The aircraft's permit-to-fly is restricted to VFR flights and this drastically reduces the ferry range as it is rare to fly at optimal altitude. In this 'low-level' regime the Sabre gulps fuel, using about 1.4 US gallons per mile at 250 knots, which gives a still-air range for planning purposes (with reserves) of about 400 miles. An added complication is that all performance manuals for the Sabre quote fuel burn in pounds.

The pneumatic system connects to anti-'g' pants to assist the older and more gentlemanly pilot with handling the 'g' forces. That said these pants are not really needed as the aeroplane is flown to a self-imposed limit of 4 'g' (the aircraft is actually stressed to 7 'g'). The intention being to respect the aircraft's age and historical importance and therefore pilots always try to fly her well inside the speed, manoeuvre and 'g' limits permitted.

Landing

Because the Sabre uses fuel so quickly, sooner than wished the pilot has to think about bringing the aircraft back in to the circuit. The circuit is usually joined from a climbing break to land which, in conjunction with the highly effective speed-brakes, helps the aircraft to slow down. The speed-brakes draw a big load from the hydraulic system and they do not always come out exactly symmetrically: if this is the case the aircraft snakes a tiny bit, but not enough to be of any concern. Once the speed-brakes are deployed fully the hydraulic pressure starts to rise again. There is no gauge to show speed-brake position, but there is no doubting that they have deployed as they can be heard and felt, plus they produce a noticeable pitch trim change and give lots of drag, requiring a big increase in power to hold the aircraft's speed.

Next it's time for the landing gear to go down and any time below the limiting speed (186 knots, conveniently marked on the Airspeed Indicator

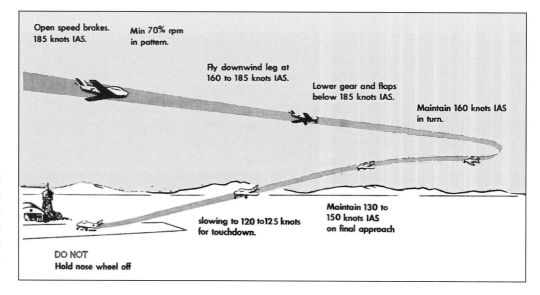

RIGHT The speeds on this diagram have been adjusted to reflect the lack of slats on 48-178, i.e. 10 knots have been added to all speeds.

Open speed brakes. 185 knots IAS.

Min 70% rpm in pattern.

Fly downwind leg at 160 to 185 knots IAS.

Lower gear and flaps below 185 knots IAS.

Maintain 160 knots IAS in turn.

Maintain 130 to 150 knots IAS on final approach

slowing to 120 to 125 knots for touchdown.

DO NOT Hold nose wheel off

with a thin yellow line) the electrically driven flaps can be lowered. The flaps are only operated in one of two positions, up or down: there is no indication of the flap position in the cockpit, but again this is not a problem as their position can easily be seen by looking out the window, plus there is a big pitch down trim change associated. The flaps provide lots of lift and lower the nose significantly on the approach; however, they do not produce much drag, which is one reason for landing with the speed-brakes out.

In this configuration (gear down, flaps down, speed-brakes out) the aircraft has enough power to easily go-around. It is aimed to keep the engine at a minimum of 70% rpm to facilitate this contingency: with engine speed above 70% the spool-up time is actually fairly respectable. Below 70% things are different: from idle power it takes about 20 seconds for the engine to deliver full power. This can, quite literally, feel like a lifetime if you really need power at that precise moment, and can be the end of a lifetime if you needed it any more than a few seconds ago. The protracted acceleration time cannot be rushed either as the engine will only surge if hurried. It's not the intention to over-sensationalise what is (by modern standards) a really appalling trait. The fact is that all the pilots understand the limitations of the J47 and have adopted the very same techniques used when in service to mitigate the protracted wind-up time. Slow engine response was a fairly typical problem with first-generation jet engines. The behaviour of this engine was unacceptable (British test pilot Duncan Simpson who evaluated the

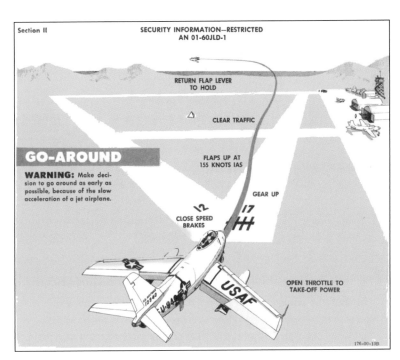

RETURN FLAP LEVER
TO HOLD

CLEAR TRAFFIC

GO-AROUND

WARNING: Make decision to go around as early as possible, because of the slow acceleration of a jet airplane.

FLAPS UP AT
155 KNOTS IAS

GEAR UP

CLOSE SPEED
BRAKES

OPEN THROTTLE TO
TAKE-OFF POWER

176-00-10B

Sabre for the RAF said as much) and it needed to be fixed. Eventually, the boffins did come up with the technical solutions needed and pilots of modern machinery enjoy a situation whereby it's virtually impossible to surge a jet engine no matter how much abuse is thrown at it.

It is aimed to tip in to the final turn at 150 knots, reducing to 130 knots once lined up with the runway. Touch down is at 125 knots at normal weights. The tendency is to land with some power still applied: if the throttle is closed too soon the landing can be 'firm' given that the nose-high attitude means that a proportion of the aircraft's weight is being supported by the engine.

ABOVE The Sabre has more than enough power to carry out a go-around with gear, flaps and speed-brakes all extended.

LEFT F-86A in landing configuration: gear down, flaps down, speed-brakes out.
(Richard Paver)

ABOVE Although 48-178 wears the Korean War markings of the 4th Fighter Interceptor Wing this aircraft never actually served in Korea. Most of the aircraft's operational career was spent on the US west coast including duties protecting the Los Alamos National Laboratory. *(Gareth Horne)*

ABOVE The nose-high landing attitude is clearly shown in this shot of the Sabre coming in to land at Biggin Hill in 2008. Touch-down speed is 125 knots, about 10 knots faster than normal due to the leading edge slats being in the fixed closed position. *(Martin Stephen)*

One area in which the Sabre excels is in her ability to stop. The brakes are incredibly good and they grip fantastically well, even in the wet or in a cross-wind. There is actually a slight negative angle of attack when the aircraft is on all three wheels and this gives racing-car spoiler-style aerodynamic down-force, thereby pushing all the weight on to the wheels and tyres. Notwithstanding this, there is no anti-skid braking system (it hadn't been invented yet) so care must still be taken not to be too heavy-footed in the wet so as to avoid aquaplaning or flat-spotting, or even bursting a tyre. There is no advantage to be had in aerodynamic braking in a Sabre, unless the runway is a long one.

The shut-down procedure is very simple and consists of simply switching everything off. Once it has all gone quiet the aircraft merrily ticks away as the engine gently cools down.

48-178 is an extraordinary aircraft and those who fly her are astonishingly fortunate. Golden Apple Trust's ambitious, challenging and dedicated undertaking to preserve the Sabre in airworthy condition is an awesome challenge. It can only be guessed what effect this aircraft has on those who see it flying, but perhaps someone somewhere will be inspired by it and be motivated in some small way to become involved in aviation. If that is the case then the Trust is fulfilling its remit and it has all been worthwhile.

For me personally there is no other legendary aircraft that combines performance, style and history in the way that the Sabre does. It is simply the coolest aircraft that I have flown, ever.

Public display sequence

The Sabre is a great aircraft to fly at public displays. It looks terrific from any angle, especially top-side, with smoke pouring from the engine adding something very unique to the spectacle.

Luckily for the pilots it's almost impossible to fly a bad show in a Sabre simply because it has such tremendous aerial presence. There is absolutely no rationale to do anything other than a series of graceful swooping manoeuvres.

BELOW Golden Apple's F-86A displaying at one of the Shuttleworth Collection's air displays at Old Warden in 2009. *(Nick Blacow)*

The F-86A is a very popular air-show aircraft and for good reason. In European skies it is unique, it looks beautiful and, with the exception of some national formation display teams, it is the smokiest aircraft on the display scene.

In the diverse world of aviation the Sabre holds a very special place and by keeping it flying it is hoped that it stirs emotions that could never be evoked by a dusty static museum piece. This is why air-show flying is a very important activity for the Golden Apple Trust. The pilots have a responsibility to show the aircraft in a dynamic environment that best demonstrates the aircraft's characteristics yet at the same time respects her age and historical importance. Displaying an aircraft like the Sabre should never be about any individual pilot showing off their skills. Fortunately for us, the F-86A is the sort of aircraft that can be flown benignly without resort to any tricks, stunts or other crowd pleasers. It can be flown with spirit and well within the flight envelope and yet still convey the notion that for its day this was a potent machine with amazing performance.

My own display sequence is very straightforward and has been developed over many years. It comes from the K-I-S-S (keep-it-simple-stupid) school of flying and by following it I ensure that there is always a big margin between where I plan to take the aircraft in the sky and any place where I might inadvertently paint myself into a corner, so to speak. Every display pilot should know that it's OK to be low on energy and it's OK to fly low, but chickens come home to roost very quickly (and tragically) if one tries to do both of these things at the same time. I never forget this basic rule. The sequence also copes with varying weather and wind conditions and means I can go flat, full or rolling as and when I see fit if conditions allow. There is no question of committing to a full (vertical) show and then getting stuck for ideas when the cloud rolls in. The sequence also allows me to fly a full show at one end of the crowd-line and flat at the other end should the weather or ATC constraints dictate.

In Sabre displays you will never see inverted flight, knife-edge flying, stalling and

ABOVE 48-178 taxies for a display at Marville (a former Canadian Air Force Sabre base) in France in 2005. *(Emmanuel Perez)*

LEFT 48-178 is the only airworthy 'A' model variant of the legendary NAA F-86 in the world and the only Sabre of any series still flying in Europe. *(Author)*

BELOW The Sabre parked outside a hardened aircraft shelter at the RNLAF Twente Air Show in 2005. Note the open ammunition bay behind the nose-wheel door. On land-aways this large storage space is put to good use to transport engine blanks, the canopy cover, tools, spare oil and the pilot's luggage (with room to spare). *(Author)*

ABOVE 48-178 after landing at Edinburgh Airport following a display at Scotland's National Museum of Flight, East Fortune, in 2004. *(Gerry Hill)*

RIGHT The Sabre's classic shape is shown to great effect by this air-to-air shot against a dark background. *(Geoffrey Lee)*

BELOW The Sabre's classic profile shown during a display at Duxford in 2008. This photograph illustrates the long-chord ailerons and the non-standard leading edge wing fences. *(Matt Clackson)*

certainly no 'flick' manoeuvres. It's all very gentlemanly and relaxed: there is no point giving the aircraft (or oneself) an unnecessary hard time. No one will appreciate anything other than smooth, graceful lines being carved in the sky by that smoke trail.

For me the display begins about an hour before take-off and ends when I'm standing back on the ground afterwards. I start my preparations by mentally getting myself in 'the zone'. The mobile is switched off and I find somewhere quiet so that I can walk through the sequence. I've flown the routine literally hundreds of times, but I still need to practise it mentally in the context of the venue and weather conditions, which are never the same twice. Getting oneself into the correct state of mind or addressing the psychological aspects as they relate to display flying are, in my opinion, just as important as having the necessary flying skills. Every display pilot will feel nervous (stage fright?) to a degree and this is not unhealthy as (in my opinion) it helps to focus the mind. There is certainly no place for complacency nor for being too 'wired' in this arena and thus mental preparation is my way of getting the balance right.

There is always a great deal of planning before a display. Other than shows at Duxford, every event requires flying the Sabre across country. Sometimes we land-away, other times we fly in and out of a slot. Many venues do not have the runway or facilities to handle an F-86 so we will operate in and out of another airfield. Often we attend more than one show on a day. All of this requires a high degree of planning and split-second execution: flying display programmes are often very busy and being even a minute late for a slot can mean inconveniencing other participants or even force a cancellation by the Display Director should the flying programme be thrown into disarray as a result. Being too early is almost as bad, especially when fuel reserves are tight, which is always.

Where possible I start the sequence with a topside pass. This gives the photographers the opportunity to capture the classic Sabre profile and shows off the aircraft's definitive shape. It would be very easy to run in on this initial pass with a fanfare at 450+ knots;

however, this speed would be excessive, leaving the aircraft with too much energy for the next manoeuvre. The best turning speed and pitch response occurs at about 330 knots: any faster and turns become very distant and protracted, therefore to keep the aircraft in sight of the crowd I try to hold back on the speed. The time for a high-speed pass will come at the end. With about 90% power and 330 knots at crowd centre a series of vertical and rolling manoeuvres can be flown comfortably yet remain fairly tight to the crowd without the need for excessive 'g' forces. Another benefit is that at this speed the aircraft is very well behaved and far removed from any of the instability problems that can affect it at either end of the spectrum. The energy is traded for height as the aircraft goes up the hill and, because the Sabre is very clean and maintains energy really well, is restored quickly on the way down. Energy management in the Sabre is usually focused on having too much, which is a nice problem to have.

For me the key to a good Sabre display is to link together a series of smooth swooping manoeuvres with some slightly more dynamic flourishes that show the aircraft's excellent roll and pitch rates. At no time is there any need for 'aggression' or grunting: fortunately, the Sabre is very good at letting you know that you are trying a little too hard. I feel very relaxed once I am in the display: it's a nice place to be, I do not have to work too hard at it and of course I'm enjoying myself, which is really how flying should be. I honestly feel more stressed navigating the M25 at rush hour on a wet Monday morning.

Towards the end of the sequence I start to nudge the power up in order to gather a few extra knots for the final flypast. I want to get as much out of the old girl as I can for this. It serves to demonstrate just how much of a quantum leap in performance this aircraft was compared to the last of the piston-engine fighters that it followed out of the factory in 1947. If the weather is good and there are no ATC restrictions I can also stand the machine on its tail and vanish into the vertical, trading all that energy for altitude in a manoeuvre that still leaves me, and hopefully the spectators, in awe.

ABOVE Golden Apple stablemates Lockheed Canadair T-33 (G-TBRD) and NAA F-86A (G-SABR) captured in formation in 2004. These aircraft regularly performed together at air shows. The T-33 had much greater range, but the Sabre's top speed was much faster. The aircraft were reasonably evenly matched in performance terms for flying in formation. Sadly, the T-Bird was destroyed in a take-off accident late in 2006; fortunately, both occupants received only minor injuries. *(Jenny Coffey)*

LEFT A practice display on a grey day at Duxford. As can be seen, the pilot's view 'over-the-shoulder' is exceptional.

BELOW A top-side pass is obligatory and the best way to show the Sabre's classic lines. *(Mike Hall)*

MiG-15 –
A worthy adversary

A description of the Mikoyan-Gurevich MiG-15 is included here to provide direct comparison with the F-86A. Both aircraft featured the same wing-sweep angle and shared similar configurations; however, it was there that the similarities ended. The aircraft used different types of engine and (in accord with the two nation's different philosophies) were armed differently: the MiG was built to take down bombers (cannon), whereas the Sabre was conceived as a pursuit/escort fighter (machine guns). Some of the handling problems associated with swept-wings had been addressed by the designers using different solutions. These dissimilarities could be, and frequently were, exploited to devastating effect.

The MiG was more heavily armed (but wore only minimal armour), lighter and had more thrust, whereas the Sabre was a much more stable weapons platform, had better handling, a superior wing and offered the pilot unmatched visibility and operability. The MiG was unquestionably less refined than the Sabre, being designed with operation from basic airfields, far from sophisticated support structures, very much in mind. It also made due allowance for maintenance being carried out by someone with the skills of a good tractor mechanic. Furthermore, the extensive use of pneumatics instead of hydraulics probably has something to do with the anticipated use of the aircraft from bases in the frozen Siberian north: a lesson learned from the Second World War.

The MiG-15 featured here (registration G-OMIG) was operated by the Old Flying Machine Company until 1996 and was also based at the Imperial War Museum, Duxford. At that time it was the only one flying in private hands in Europe. In the 1990s I was fortunate to fly her on the display circuit paired with the F-86A in a Korean War demonstration. Invariably, I would fly the MiG with Mark Hanna piloting the Sabre; in these scenarios the script required that the Sabre would come off as the victor.

Licence-built as a single-seat MiG-15 in Czechoslovakia in 1953, G-OMIG was exported to Poland in 1955 where later it underwent local conversion to the two-seat SBLim-2Art specification for use as an artillery spotter aircraft. The rear seat had no flying controls, but it was equipped with basic flight instruments and a radio installation intended for use by the artillery reconnaissance officer. The aircraft retained two 23mm cannon of the original and had an RD-45F (Soviet copy of the Rolls-Royce Nene engine) installed. (The UK government had ludicrously 'gifted' this engine to the Soviet Union after the Second World War when Russia was still seen as an ally. In hindsight it was probably unwise for the British, who at that time led the world in gas-turbine technology, to hand over her secrets so freely. Stalin was so surprised at the 'gift' that he thought it must be part of some bizarre British subterfuge. This act of political naivety and industrial irresponsibility allowed the Russians to short cut what would otherwise have been a protracted development

process for their own gas-turbine engine.)

The MiG-15 was built to last: almost everything about it feels over-engineered. The flying controls, in particular, are reassuringly solid. It is also surprisingly roomy, although the pilot does sit considerably lower down than in the Sabre with nowhere near as good a view of the outside world. The visibility from the MiG's cockpit may well be poor by Sabre standards, but it is the usability therein that is really a defining difference between the two. There is no question in my mind that the MiG must have been a much harder aircraft to operate in poor weather, at night or especially in combat.

As mentioned earlier, the Sabre has a very ergonomic cockpit and this certainly makes it very easy to operate. In contrast, the MiG-15's controls and switches are scattered about in a seemingly random fashion. For example, it's necessary to search along the left cockpit wall to find the speed-brake control lever and the pitch-trim switches, neither of which fall easily to hand or eye. That said, there are many things about the MiG that are good examples of form following function. I once asked a Polish engineer, who was very experienced on the MiG, why the pneumatic transfer valves were so large. He explained that they were designed to be operated by a pilot wearing thick mittens when flying from Siberian airfields in winter. This says a lot about the thinking behind the aircraft.

Early centrifugal compressor engines demand correct handling and (just as with the Sabre) the throttle on the MiG cannot be slammed about with impunity: the RD-45F (like

the J47) is totally unforgiving of any misuse. To achieve take-off power from idle the throttle is slowly opened up to 8,000rpm; a pause to check temperatures and pressures; then brakes are released and the throttle moved more swiftly to full power at 11,560rpm. The initial acceleration is much brisker than the F-86A and in just a few seconds the needle on the airspeed indicator, which starts reading at 70 knots, is advancing. The nose is quite light and the engine thrust tends to make the nose rise. This has to be controlled positively to stop the tendency for the aircraft to over-rotate and therefore a bit of forward pressure on the stick is required until take-off speed is achieved. Unlike the Sabre's slightly nose-down attitude

ABOVE G-OMIG in one of the public hangars at Duxford in the mid-1990s. This aircraft was later traded to the Sao Carlos–Francisco Pereira Lopes (SDSC) museum in Brazil. *(Jim Groom)*

FAR LEFT AND LEFT The MiG-15's cockpit was 'agricultural' and chaotic. In terms of visibility and usability it was in a different and lesser league to the F-86.

on the ground, the MiG sits nose up and this might explain why it leaps off the ground without any encouragement at 125 knots.

The MiG-15 is very sensitive and twitchy in roll at low speeds and the ailerons must be checked neutral during take-off. To help with this a vertical white line is painted on the instrument panel to help the pilot keep the control column in the middle. The drop tanks seem to damp out the jitteriness of the roll control slightly compared to how it feels with a clean wing.

After take-off the clean-up sequence is awkward and things happen quickly. Landing gear master ON; gear selector UP; lights out; three reds; gear selector NEUTRAL; gear master OFF; then at a minimum of 150 knots, flaps UP; visual check; flaps NEUTRAL; and finally power back to the climb thrust of 10,870rpm. By the time all of this has been accomplished the aircraft is passing 200 knots and accelerating quickly.

In the air the MiG-15 has a solid feel to it. In pitch it is light (too light in my opinion) and responsive. The MiG is highly susceptible to 'dig in' or 'g' overshoot (as previously described) to the extent that some forward stick pressure is needed in a sustained high 'g' turn. In extreme manoeuvres the aircraft can easily be overstressed or depart from controlled flight. In contrast, the F-86A is nowhere near as

inclined to exhibit this unstable and potentially deadly handling characteristic. In roll the control forces are heavier than the Sabre, and although more responsive the overall rate of roll that can be achieved is slower; these roll characteristics are probably down to the MiG's ailerons being much smaller than those of the F-86. The MiG's control column is twice as tall as the Sabre's and therefore must be moved twice as far to get full deflection. This makes for the odd sensation that the aircraft is being flown with a broom handle. Like the Sabre the pitch, yaw and roll control is manual (ailerons are boosted); however, it does not have the Sabre's pitch boost, which to be fair it doesn't need given the lightness of this control. This aileron boost is not as powerful as the Sabre's, but then it is definitely not as unsubtle either. Just as with the F-86A, a failure of the aileron boost would require both hands to get the aircraft to roll at a respectable rate, and probably not at all at high speed. Interestingly, early MiG-15s did not have any aileron boost, which might also explain the tall control column to give a longer lever arm to permit any meaningful movement at high speed.

The MiG's 'g' limits are very similar to the F-86's, although the MiG pilot did not benefit from an anti-'g' suit. Most vertical manoeuvres (loops, barrel rolls and half cuban eights) are flown comfortably from a starting speed of 320 knots. The only unusual characteristic during aerobatics is a tendency to sideslip when going slow over the top of a manoeuvre. The rudder forces are light, however, and the aircraft is easily straightened. There is a tuft of wool on a piece of wire in front of the windscreen to show sideslip, which is simple yet effective. In my opinion the MiG-15 is a bit too short, which might explain why directional stability is so weak. Indeed, the aircraft at all speeds seems to continually 'wag' its tail and in combat it must have been difficult for the pilot to find a steady tracking solution for the cannon. The addition of a second cockpit (on G-OMIG) probably makes this trait worse, but the general deficiency was addressed on the MiG-17 by, among other changes, lengthening the aircraft, thereby giving the fin a greater stabilising moment.

At very high speed, due to a combination of adverse aileron yaw, combined with weak directional stability and poor torsional wing stiffness (whereby the high air loads on the ailerons caused the wing to twist adversely), the MiG-15 was known to exhibit roll-control reversal. This highly dangerous situation caused the aircraft to turn in the opposite direction to the one the pilot intended. Furthermore, shock waves acting on the elevators at very high Mach number could cause control reversal in pitch too. The consequences of this trait are known to have caused structural failure (due to over-'g'), departure from controlled flight and many a 'bent' airframe/frightened pilot. I never experienced this phenomenon myself due to not flying faster than the drop tank speed limit of 430 knots, although above 400 knots the ailerons do become excessively heavy and things start to feel a bit odd, as if the machine is warning you not to go any faster.

The Russian fuel system is not as well thought out as the American equivalent. The only gauge in the cockpit is one that indicates the contents of the No. 2 (front) tank (700 litres) and when that starts to move down it is time to find somewhere to land. Fuel feeds from the drop tanks first then from the No. 3 (rear) tank to the No. 2. When the No. 3 tank empties a small light illuminates on the instrument panel to remind you to switch the booster pump off: failure to do this will cause the pump to run dry and burn out (apparently). When the No. 3 tank light comes on you know that you have 700

litres remaining. This makes the pilot's life more difficult than necessary: 700 litres is not much in the MiG and the light coming on is the first real indication of the fuel state. Up until then it's all been mental dead-reckoning. Fuel 'hanging-up' in one of the drop tanks (not uncommon) is indicated by an increasing amount of aileron trim required to fly straight.

Weak directional stability makes the MiG somewhat prone to spinning at low speed. With unfortunate stall characteristics (and a lack of any real stall warning) departure from controlled flight would have been an ever-present danger for the unwary Fagot pilot, especially in the heat of battle. Other deficiencies identified in post-war US testing of a MiG-15 revealed poor lateral and directional stability at high speed and high altitude, as well as an inferior rate of roll. These factors meant that the MiG was never able to offer the pilot the kind of 'carefree' handling that is (almost) a characteristic of the Sabre: the F-86 is not perfect but is still far more tolerant of vigorous control inputs than the MiG-15. The Sabre pilot needs really to push the aircraft around carelessly to get it to flick into a spin; not so for his Soviet counterpart.

At pattern speed the MiG is a bit of a handful. The pitch trim is electric with the control inconveniently mounted out of the way on the left cockpit wall. This is unhelpful given the large adjustments in trim that are necessary when lowering the gear and flaps. During the final approach, with 50 degrees of flap and the speed back at 130 knots, the MiG starts to show its boundaries in several ways. In addition to poor handling, slow engine response and abrupt drag curve issues associated with many early swept-wing aircraft (Sabre included), the MiG has an additional problem (which the F-86A most certainly does not have) in that it possesses a tendency to become unstable in both roll and yaw. The onset of this condition (commonly known as 'Dutch Roll') is most apparent in the landing configuration. Dutch Roll is a type of motion that can be described as an out-of-phase combination of 'tail-wagging' and 'wing rocking'. Dutch Roll results from the relatively weak directional stability of the aircraft when compared to its strong lateral stability. Any sideslip on the approach causes roll as a secondary effect and the aircraft's lateral stability then tries to restore itself to level flight. At the same time, the weaker directional stability attempts but fails to correct the sideslip. Since directional stability is weaker than lateral stability the restoring yaw motion lags significantly behind the restoring roll motion. As such, the aircraft passes through level flight as the yawing motion is continuing in the direction of the original roll. At that point, the sideslip is introduced in the opposite direction and the process is reversed. In the MiG-15 this roll and yaw motion steadily increases in amplitude with a frequency of 1 or 2 seconds. Any attempt by the pilot to counter this motion usually only makes the problem worse as it is easy to get into a Pilot Induced Oscillation (PIO). The Dutch Roll is manageable if no attempt to correct it is made: in other words you can live with it, but it is somewhat disconcerting. The only fix if the motion becomes severe is to execute a go-around or, if time and height allow, to let go of the stick to let the aircraft calm itself down. Better still the condition can be avoided by not making any abrupt rudder inputs so as to reduce the chance of any sideslip developing. On landing, the Dutch Roll causes the aircraft to hop from wheel to wheel for a few cycles before settling. This is all quite alarming the first time it's experienced and evidence again (if needed) that the MiG-15 was a little too short or could have done with a slightly larger fin to give it stronger directional stability. Second-generation

swept-wing aircraft incorporated yaw dampers specifically to deal with any inclination toward Dutch Roll.

The greatest apprehension about landing the MiG concerns its ability to stop. Provided the aircraft is landed in the right place and at the correct speed (125 knots), all will (in theory) be well. There is a temptation to apply the brakes too early or too harshly on a short runway, which only serves to lock up the wheels (there is no anti-skid). The aircraft is easy to keep straight but because of the slightly nose-high attitude when on the ground it takes a while for all the lift to vanish and for the weight to settle on to the wheels. Only then can any effective braking be carried out. Like the Sabre, and for the same reason, the landing is carried out with the (less effective) speed-brakes out.

In summary, the MiG-15 is a sturdily engineered little fighter. Engine handling and flight on the landing approach demand some care, but it has good performance and, like the Sabre, is very enjoyable to fly in the middle of its envelope. At the edges it must be flown with more restraint. Respect for its age and some caution about its early design status is wise, given that it is not as carefree an aircraft as we have come to expect from modern jet-fighters. Any aircraft will 'bite', but the MiG-15 has the potential to do this very rapidly, especially in pitch during high-energy manoeuvres.

Would I prefer to fly the F-86A or the MiG-15? In war or in peace for me it's got to be the Sabre. Its overall performance is not as good as the lighter and more powerful Russian. Its handling at low speed is not brilliant, but it is far better than the MiG's. The fact that it is a much more stable gun platform counts for a lot, not least in combat. For me, the Sabre is the more iconic, historic and handsome of the two. Performance, speed and handling are important, of course, but all aircraft have subjective qualities that are more difficult to label or quantify. Perhaps when all is said and done I would choose the Sabre over the MiG simply because I feel more 'in tune' with it. The Sabre was a style icon of its time and was once described as 'a machine of almost lyrical beauty'. Also, and not inconsequentially, I think the F-86 is somehow 'cooler' than any comparable aircraft.

BELOW The Old Flying Machine Company's MiG-15 landing at Woodford in 1995. The speed-brakes were smaller and much less effective than the Sabre's. In this picture the nose-up stance when on the ground is also evident.
(Chris Muir)

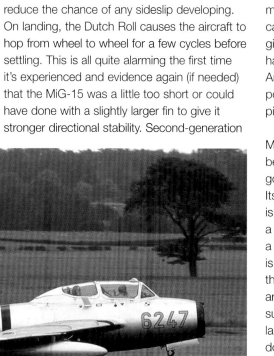

Comparison of Sabre (F-86A) and MiG-15

Specifications	F-86A Sabre	MiG-15 Fagot
Wingspan	35ft 11in	33ft 1½in
Wing area	287.9sq ft	221.7sq ft
Length	37ft 6in	33ft 3⅝in
Height	14ft 8in	11ft 2in
Max. take-off weight	13,791lb	11,270lb
Armament	Six 0.50-calibre machine guns and sixteen 5in rockets or 2,000lb of bombs	Two 23mm cannons and one 37mm cannon, plus rockets or 2,000lb of bombs
Engine	One General Electric J47 engine with 5,200lb of thrust	One VK-1 (RD-45F) of 6,000lb thrust (copy of British Rolls-Royce Nene engine)
Power to weight ratio	0.377	0.532
Max. speed	679mph	670mph
Cruising speed	540mph	525mph
Range	1,052 miles	500 miles
Service ceiling	51,000ft	55,000ft
Time to climb to 45,000ft	13 minutes	9 minutes
Equipment	Anti-'g' suit	No anti-'g' suit
	Defensive armour around the cockpit – forward and aft, plus bullet-proof front screen and guns	Defensive armour around the cockpit – aft only, plus bullet-proof front screen
Internal heating and defrosting	Excellent and effective	Deficient and ineffective, particularly at high altitude, where the machine enjoyed its principal performance advantage. There are accounts of MiGs descending to engage with Sabres, but with badly frosted canopies
Gun sight	A-1CM radar gun sight with AN/APG-30 ranging radar	ASP-3N, a crude optical gun sight without ranging radar

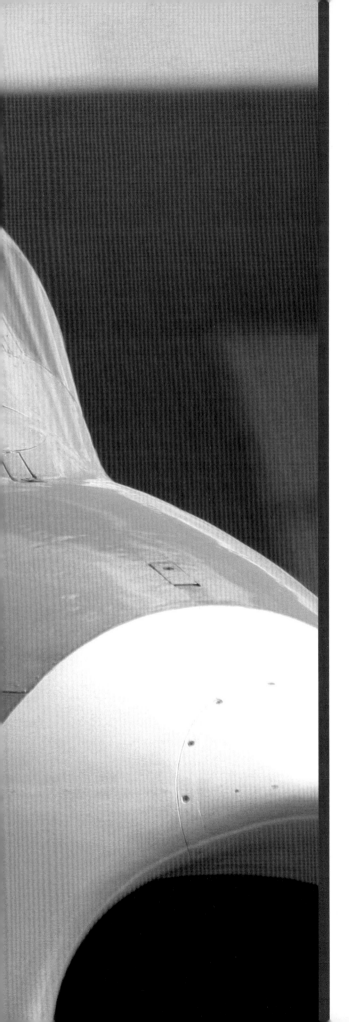

Chapter Five

Anatomy of the Sabre

In its day the F-86 was a completely new breed of aircraft: its performance, engineering and technology were in an entirely different class to any that had gone before. The aerodynamicists had their own problems to solve, but it fell to the structural designers and manufacturing engineers to turn those innovations into metal.

OPPOSITE As pioneers of new structures, systems and the aerodynamics of high-speed flight the creators of the Sabre faced many challenges. The elegance of the result is a tribute to their skill: the aircraft was both an engineering triumph and a design classic.

TABLE OF CONTENTS

The aircraft's structure would have to withstand stresses and strains that were far greater than any piston-powered fighter or early straight-wing jet. Many of the aircraft's ancillary systems would also need to be clean-sheet designs to cope with the demands of a much expanded flight envelope. The Sabre had to be light, it needed to be easy to maintain and it had to be very, very strong. From a commercial perspective it had to be manufactured on an industrial scale and yet still remain affordable. It is to the enormous credit of the men and women of NAA that all of these sometimes conflicting requirements were achieved.

Author's note: the following section contains highly abridged descriptions taken from the original flight manual as well as contemporary unofficial descriptions of the aircraft's systems. Please refer to USAF publication AN 01-60JLA-1 for a more in-depth explanation of the aircraft's systems, their correct use and safe operation. This chapter is intended only to give the most general overview of the aircraft's operation and systems.

General description

The NAA F-86A Sabre is a single-seat, high-altitude fighter, powered by an axial-flow turbo-jet engine. It is characterised by swept-back wings and empennage. Noteworthy features

ABOVE The contents page from the 1953 Flight Operating Instructions for the F-86A. Many operations manuals of the era were laced with cartoons and illustrations intended to catch the eye, convey vital information and keep the reader interested. Modern aircraft manuals show no such humour or style.

RIGHT F-86A general arrangement.

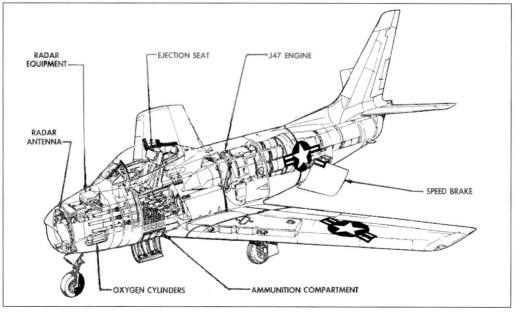

RADAR EQUIPMENT — EJECTION SEAT — J47 ENGINE — RADAR ANTENNA — SPEED BRAKE — OXYGEN CYLINDERS — AMMUNITION COMPARTMENT

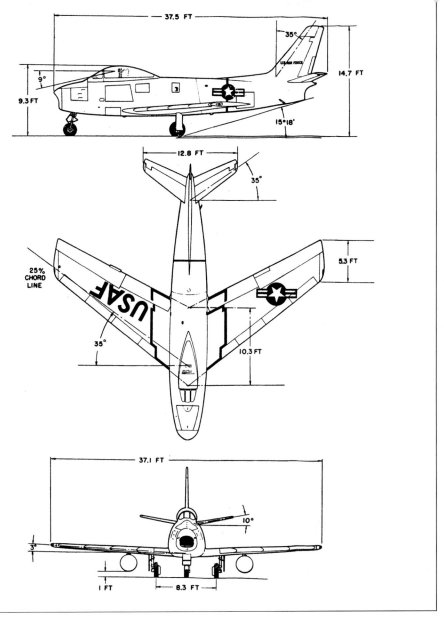

include an adjustable horizontal stabiliser, leading
edge wing slats and fuselage speed-brakes.
The aircraft is designed primarily for flight in the
transonic speed range at high altitude. The aircraft
may also be used to attack ground or naval
targets with gunfire, bombs, rockets or chemicals.

Armaments – guns,
rockets, missiles and
bombs

All 'A' model aircraft and most other
subsequent Sabres were armed with six
0.50-calibre Type M-3 machine guns mounted
in the forward fuselage, three on each side of
the engine air intake duct. It was also capable
of carrying up to 2,000lb of bombs, rockets

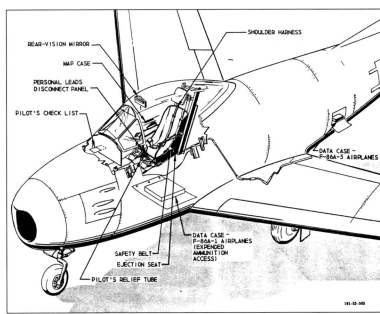

REAR-VISION MIRROR

MAP CASE

PERSONAL LEADS
DISCONNECT PANEL

PILOT'S CHECK LIST

SHOULDER HARNESS

DATA CASE –
F-86A-5 AIRPLANES

DATA CASE –
F-86A-1 AIRPLANES
(EXPENDED
AMMUNITION
ACCESS)

SAFETY BELT

EJECTION SEAT

PILOT'S RELIEF TUBE

101-53-565

North American

F-86A Sabre cutaway

drawing. *(Mike*

Badrocke)

1　Radome
2　Radar antenna
3　Engine air intake
4　Gun camera
5　Nose wheel leg doors
6　Nose undercarriage leg strut
7　Nose wheel
8　Torque scissor links
9　Steering control valve
10　Nose undercarriage pivot fixing
11　Sight amplifier
12　Radio and electronics equipment bay
13　Electronics bay access panel
14　Battery
15　Gun muzzle blast troughs
16　Oxygen bottles
17　Nose wheel bay door
18　Oxygen servicing point
19　Canopy switches
20　Machine gun barrel mountings
21　Hydraulic system test connections
22　Radio transmitter
23　Cockpit armoured bulkhead
24　Windscreen panels
25　A-1CM radar gun sight
26　Instrument panel shroud
27　Instrument panel
28　Control column
29　Kick-in boarding step
30　Used cartridge case collector box
31　Ammunition boxes (267 rounds per gun)
32　Ammunition feed chutes
33　0.5in (12.7mm) Colt Browning machine guns
34　Engine throttle
35　Starboard side console panel
36　North American ejection seat
37　Rear view mirror
38　Sliding cockpit canopy cover
39　Ejection seat headrest
40　ADF sense aerials
41　Pilot's back armour
42　Ejection seat guide rails
43　Canopy handle
44　Cockpit pressure valves
45　Armoured side panels

46　Tail plane trim actuator
47　Fuselage/front spar main frame
48　Forward fuselage fuel tank (total internal fuel capacity 434.4 US gal/1644 litres)
49　Fuselage lower longeron
50　Intake trunking
51　Rear radio and electronics bay
52　Canopy emergency release handle
53　ADF loop aerial
54　Cockpit pressure relief valve
55　Starboard wing fuel tank
56　Leading edge slat guide rails
57　Starboard automatic leading edge slat, open
58　Cable drive to aileron actuator
59　Pitot tube
60　Starboard navigation light
61　Wing-tip fairing
62　Starboard aileron
63　Aileron hydraulic control unit
64　Aileron balance
65　Starboard slotted flap, down position
66　Flap guide rail
67　Upward identification light

68　Air conditioning plant
69　Intake fairing starter/generator
70　Fuselage/rear spar main frame
71　Hydraulic system reservoirs
72　Longeron/main frame joints
73　Fuel filter de-icing fluid tank
74　Cooling air outlet
75　Engine equipment access panel
76　Heat exchanger exhaust duct
77　Engine suspension links
78　Fuselage skin plating
79　Engine withdrawal rail
80　Starboard side oil tank (5.7 US gal/21.6 litres)
81　General Electric J47-GE-27 turbojet
82　Bleed air system primary heat exchanger
83　Ground power connections

84　Fuel filler cap
85　Fuselage break point sloping frame (engine removal)
86　Upper longeron joint
87　Engine bay cooling air duct
88　Cooling air outlet
89　Engine firewall bulkhead

90　Engine flame cans
91　Rear fuselage framing
92　Fuel jettison pipe
93　Fuselage top longeron
94　Fin/tail plane root fillet fairing
95　Control cable duct

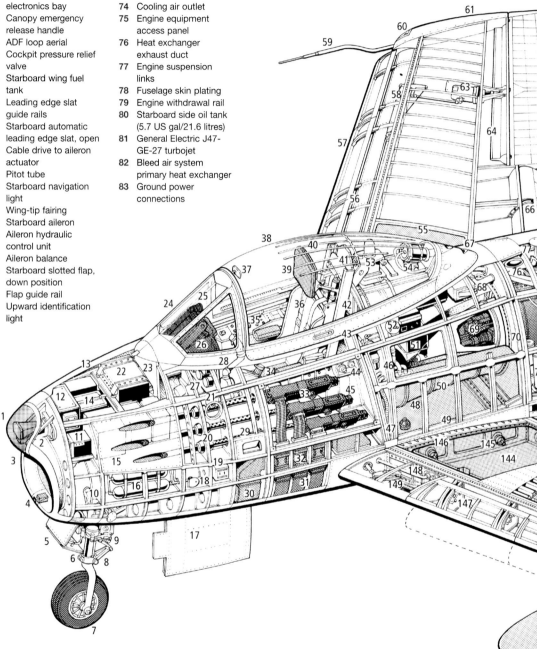

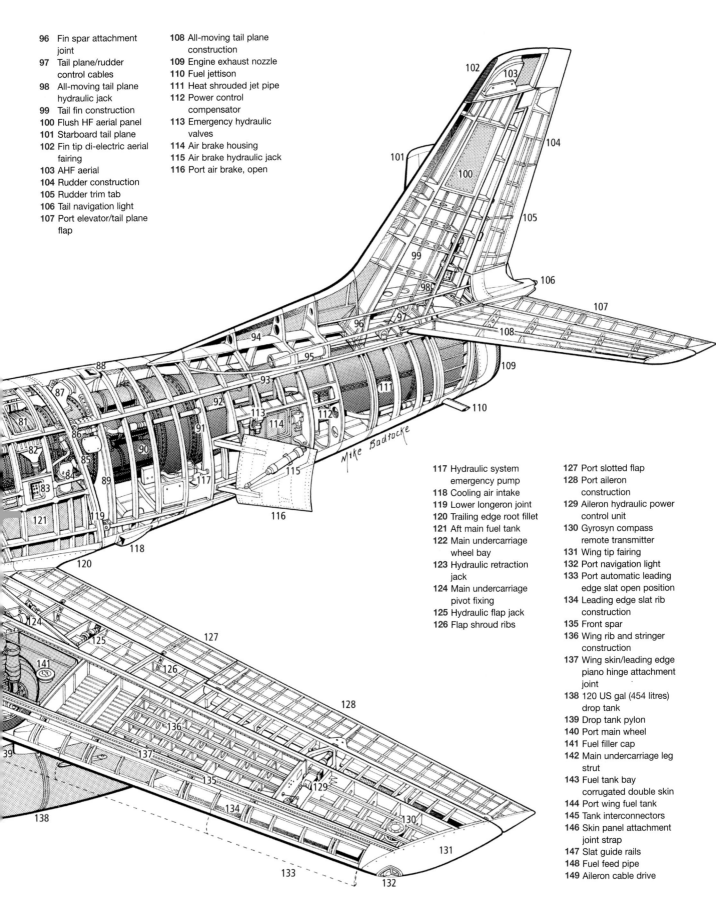

96 Fin spar attachment
   joint
97 Tail plane/rudder
   control cables
98 All-moving tail plane
   hydraulic jack
99 Tail fin construction
100 Flush HF aerial panel
101 Starboard tail plane
102 Fin tip di-electric aerial
   fairing
103 AHF aerial
104 Rudder construction
105 Rudder trim tab
106 Tail navigation light
107 Port elevator/tail plane
   flap

108 All-moving tail plane
   construction
109 Engine exhaust nozzle
110 Fuel jettison
111 Heat shrouded jet pipe
112 Power control
   compensator
113 Emergency hydraulic
   valves
114 Air brake housing
115 Air brake hydraulic jack
116 Port air brake, open

117 Hydraulic system
   emergency pump
118 Cooling air intake
119 Lower longeron joint
120 Trailing edge root fillet
121 Aft main fuel tank
122 Main undercarriage
   wheel bay
123 Hydraulic retraction
   jack
124 Main undercarriage
   pivot fixing
125 Hydraulic flap jack
126 Flap shroud ribs

127 Port slotted flap
128 Port aileron
   construction
129 Aileron hydraulic power
   control unit
130 Gyrosyn compass
   remote transmitter
131 Wing tip fairing
132 Port navigation light
133 Port automatic leading
   edge slat open position
134 Leading edge slat rib
   construction
135 Front spar
136 Wing rib and stringer
   construction
137 Wing skin/leading edge
   piano hinge attachment
   joint
138 120 US gal (454 litres)
   drop tank
139 Drop tank pylon
140 Port main wheel
141 Fuel filler cap
142 Main undercarriage leg
   strut
143 Fuel tank bay
   corrugated double skin
144 Port wing fuel tank
145 Tank interconnectors
146 Skin panel attachment
   joint strap
147 Slat guide rails
148 Fuel feed pipe
149 Aileron cable drive

Mike Badrocke

RIGHT The F-86 could
carry a large number
of different weapon
types. *(USAF)*

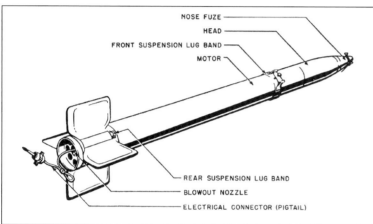

NOSE FUZE

HEAD

FRONT SUSPENSION LUG BAND

MOTOR

REAR SUSPENSION LUG BAND

BLOWOUT NOZZLE

ELECTRICAL CONNECTOR (PIGTAIL)

RESET

ROCKET
TO BE
FIRED

17

TURN KNOB
COUNTERCLOCKWISE
TO STOP

(Author)

(USAF)

ABOVE, LEFT AND BELOW Without drop tanks fitted the Sabre could carry
up to sixteen 5in High-Velocity Aircraft Rockets (HVARs).

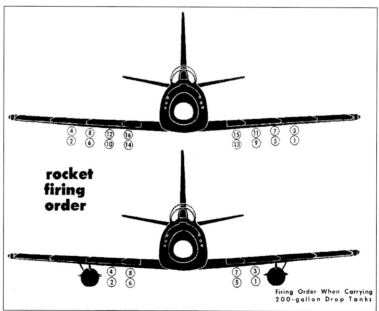

**rocket
firing
order**

Firing Order When Carrying
200-gallon Drop Tanks

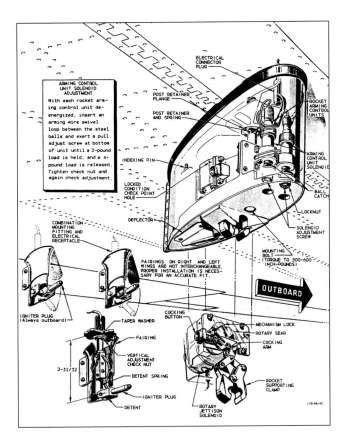

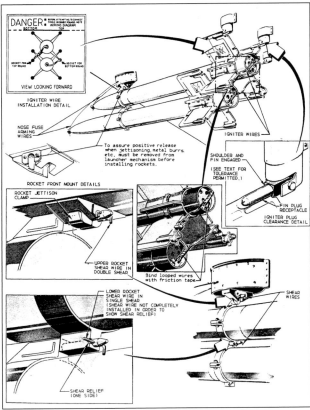

ABOVE LEFT AND RIGHT Rocket-launcher installation and HVAR loading instructions.

RIGHT After loading HVAR rockets the bore sighting was a rudimentary affair done by eye.

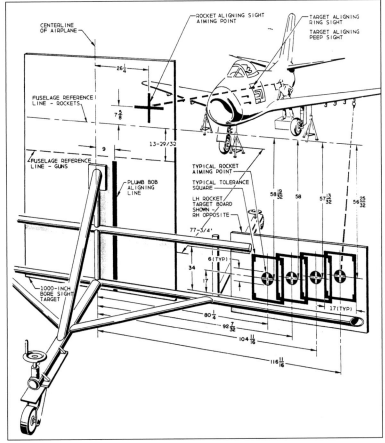

and chemicals on under-wing hard points when external fuel tanks were not fitted. Later Sabres could launch AIM-9 air-to-air missiles or drop nuclear weapons. Some F-86Fs plus the last of the F-86Hs produced were fitted with four 20mm cannon in place of the six machine guns and both sub-types could have additional under-wing hard points.

The Sabre's machine guns were electrically charged with a rate of fire of 1,100 rounds per minute and each had removable ammunition containers to hold a maximum of 267 rounds per gun; when not loaded with ammunition this bay served as a baggage storage area useful for cross-country or ferry flights. The ammunition bay door also doubled as a step to help the pilot climb up into the cockpit.

Although gun stoppages could not be cleared while flying, heaters were provided in the gun

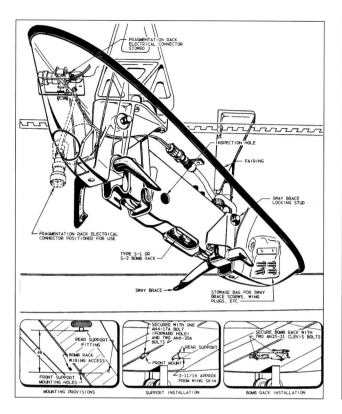

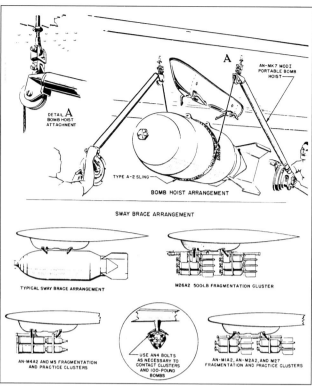

ABOVE LEFT Bomb-rack installation details.

ABOVE RIGHT Bomb-load and hoisting instructions for conventional, practice and fragmentation cluster bombs.

compartments to prevent the working parts from freezing. Additionally, air cooling was provided to the weapons during operation. Containers were fitted to the lower fuselage to collect the spent shell cases and links. A camera in the forward fuselage or nose ring (dependent on variant) ran automatically when the guns were fired. The Sabre's nose intake arrangement meant that exhaust gases from the gun barrels were much less likely to enter the engine and thereby increase the risk of engine surge; this was a problem that blighted other contemporary types with guns or cannon upstream of the engine air intake, notably the British Hawker Hunter.

All F-86A-1s and early A-5s were fitted with the Mk 18 gun sight, which projected an aiming reticule on to a reflector glass. This electrically powered gun sight was originally designed for use in bomber power-operated turrets. The sight displayed a variable-diameter circle of six diamond-shaped images surrounding a central dot and a fixed cross to show calculated lead angle. When the target was properly framed in the circle the sight would automatically compute the amount of 'lead' required. The Mk 18 was generally unreliable and was soon replaced by the more sophisticated and much improved Type A-1 CM gun sight.

FAR LEFT AND LEFT Early F-86As had troublesome flush-fitting gun doors that were designed to open in a fraction of a second when the guns were fired, but sometimes did not. These guns doors were soon replaced with open gun-barrel slots. Expendable rubber plugs were fitted over the ends of the barrels to protect them from the worst of the weather. The plugs were blown out by the first round fired.

(Author)

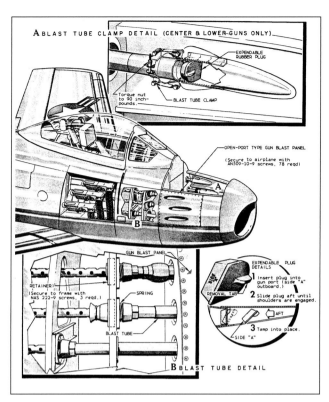

ABOVE The 0.5in Browning machine guns were installed neatly in a group of three on each side of the nose. This is the right-hand side fitment looking forward showing the metal ammunition-belt feed system. An added bonus of this location was that the guns, barrels and heavy mounting equipment also served as effective flak protection for the pilot. By contrast, the MiG-15 had very little purpose-designed or windfall armour protection. The MiG's cannons were installed much lower down the fuselage beneath the pilot's feet.

ABOVE The later type 'open' gun ports used expendable rubber plugs to protect the gun barrels from the elements.

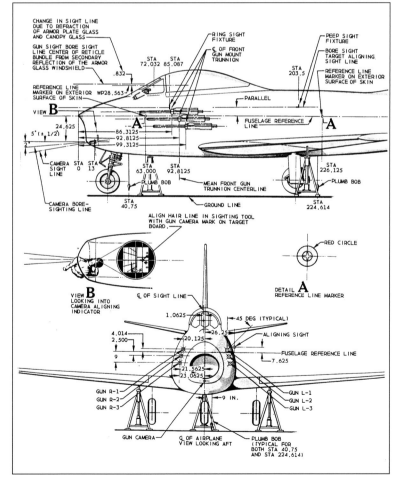

LEFT AND BELOW Harmonisation of guns, the sight and the gun camera was a protracted affair that required the aircraft to be levelled on jacks. Correct alignment of the airframe was confirmed by plumb bobs.

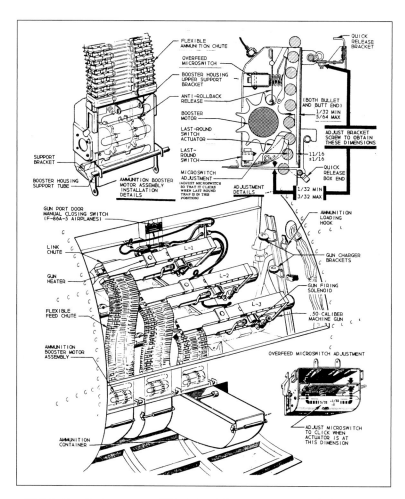

FLEXIBLE AMMUNITION CHUTE
OVERFEED MICROSWITCH
BOOSTER HOUSING UPPER SUPPORT BRACKET
ANTI-ROLLBACK RELEASE
BOOSTER MOTOR
LAST-ROUND SWITCH ACTUATOR
LAST-ROUND SWITCH
SUPPORT BRACKET
BOOSTER HOUSING SUPPORT TUBE
AMMUNITION BOOSTER MOTOR ASSEMBLY INSTALLATION DETAILS
MICROSWITCH ADJUSTMENT (ADJUST MICROSWITCH SO THAT IT CLICKS WHEN LAST ROUND TRAP IS IN THIS POSITION)
ADJUSTMENT DETAILS
QUICK RELEASE BRACKET
(BOTH BULLET AND BUTT END) 1/32 MIN 5/64 MAX
ADJUST BRACKET SCREW TO OBTAIN THESE DIMENSIONS
11/16 ±1/16
QUICK RELEASE BOX END
1/32 MIN 3/32 MAX
GUN PORT DOOR MANUAL CLOSING SWITCH (F-86A-5 AIRPLANES)
LINK CHUTE
GUN HEATER
FLEXIBLE FEED CHUTE
AMMUNITION BOOSTER MOTOR ASSEMBLY
AMMUNITION CONTAINER
L-1
L-2
L-3
AMMUNITION LOADING HOOK
GUN CHARGER BRACKETS
GUN FIRING SOLENOID
.50 CALIBER MACHINE GUN
OVERFEED MICROSWITCH ADJUSTMENT
ADJUST MICROSWITCH TO CLICK WHEN ACTUATOR IS AT THIS DIMENSION

ABOVE Detailed view of the left-hand gun installation.

This later sight was an automatic lead and ballistic computing sight used for gun and rocket firing plus an air-to-ground bomb release function. It was operable as a gun sight from sea level to 50,000ft, and as a bomb sight from sea level to 10,000ft.

RIGHT The Mk 18 gun sight proved to be unreliable in service and was replaced by the more dependable and sophisticated type A-1 CM sight.

The A-1 CM sight projected an image on to the bullet-resistant windshield glass. The image displayed the necessary lead for the machine guns to be aimed and fired; in the bombing/rocket mode it could accept manual input of wind speed and direction to allow for basic aiming corrections. F-86A-5s had manual range control, but A-6s used the AN/APG-5C and A-7s were fitted with the AN/APG-30 ranging radar. Both radars were provided to improve the accuracy of the lead angle computation; however, below about 6,000ft ranging was erratic due to ground clutter. The pilot was required to dial in the wingspan of the opposing aircraft (a MiG-15 was 33ft), close until the target indicator light glowed, smoothly track the target with the 'dot' and then fire. The control panel for sight and ranging radar was located below the instrument panel between the pilot's legs.

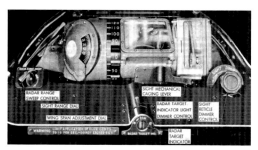

RIGHT The type A-1 CM as fitted to later F-86As, 'E's and some 'F's was an improvement over the Mk 18. The gun-sight control panel is located between the pilot's legs below the main instrument panel in front of the control column.

ABOVE Golden Apple's F-86A showing the 'vee' windscreen (a feature of F-86As (from -5 standard onward) and some early F-86Es), the six 0.5in machine-gun ports and the distinctive engine intake. *(Author)*

BELOW The Sabre pilot's operational interaction with the armament system was primarily through the control column and throttle. This arrangement pre-dated modern HOTAS (Hands On Throttle and Stick) systems by many years. Bombs and rockets were released by pressing a button on the top of the stick; the guns were fired by a trigger behind the handgrip; and the radar controlled by a thumb switch, all of which fell easily to hand. The gun-sight caging control was located on the throttle and, on early aircraft fitted with the Mk 18 gun sight, manual ranging control could be adjusted by twisting the throttle control.

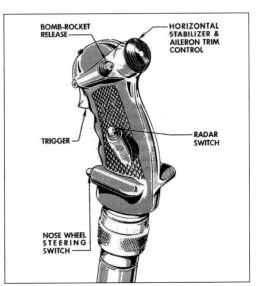

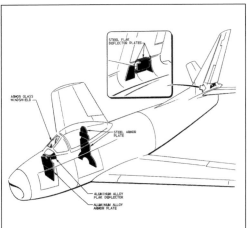

LEFT The Sabre was fitted with armour plating to protect the pilot and other vulnerable areas. Aluminium plate was fitted on the bulkhead between the battery compartment in the nose and cockpit, and around the forward cockpit instrument panel. Steel plate was welded to the rear cockpit bulkhead with an additional hinged steel plate bolted to the back of the ejection-seat guide rails to protect the pilot's head. It is doubtful that this armour would have been effective in stopping a direct hit from a 23mm or 27mm cannon shell fired by the MiG-15; however, it would have probably been sufficient to protect against small-arms fire and flak. For the F-86F further armour was fitted to protect vulnerable control linkages around the tail-plane. Of course the guns mounted on the cockpit sides were also very effective as flak protection devices. *(Author)*

range of versions before production ceased in 1956. The engine was produced in at least 17 different series and in addition to the Sabre/Fury it was used to power USAF aircraft such as the B-45 Tornado (4), B-47 Stratojet (6), Convair B-36 (4), KB-50J Superfortress (2), KC-97L Stratotanker (2), Martin XB-51 (3) and XF-91 Thunderceptor (1). Furthermore, the J47 was used to power the first land-speed record car 'Spirit of America' and the experimental M-497 Black Beetle jet-powered locomotive. Although never used as such, the J47 was also the first gas-turbine approved for civilian use.

Demand for the engine soared during the Korean War and J47 production peaked at an incredible rate of 975 engines per month. The engine saw continued service in the US military until 1978.

The following refers to the J47-GE-7 turbo-jet as fitted to the F-86A-7.

The engine is an axial-flow, jet-propulsion unit having a rated thrust of 5,200lb. In a time when centrifugal compressors were normal this arrangement was very advanced, although by modern standards somewhat unrefined. In operation air enters the intake duct in the nose and passes through an axial-flow compressor where air is compressed progressively in 12 stages. This compressed air then flows to the combustion chambers where atomised fuel is injected and combustion occurs. From the combustion chambers the hot exhaust gas

Engine

The General Electric J47 turbo-jet (GE company designation TG-190) was developed by General Electric from the earlier J35 engine and was first flight tested in May 1948. Most of the nearly 10,000 F-86s built were fitted with a variant of this engine. In all over 37,000 units were made across the full

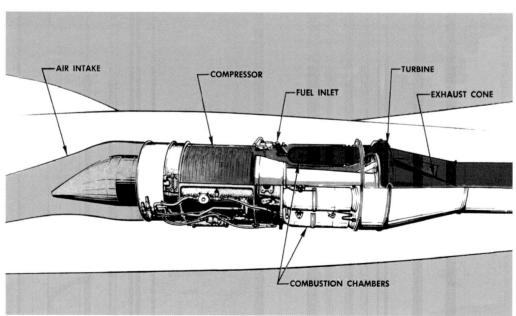

(which is expanding rapidly) passes through a turbine and on out through the tailpipe to provide the high-velocity jet and thus reaction thrust. The turbine, which is rotated by the exhaust gas passing through it, is directly connected to, and drives, the compressor stages.

When compared to contemporary piston engines, which have literally hundreds of

RIGHT AND BELOW Accessories such as the fuel pumps (normal and emergency), starter generator, inverter, oil pumps (pressure and scavenge) and the hydraulic pump are all attached to the front of the engine and are driven by gears that mate with the engine turbine/compressor shaft. A cone-shaped fairing (removed in this picture) covers these components.

(Author)

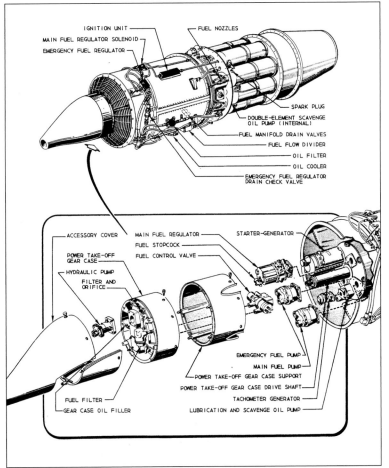

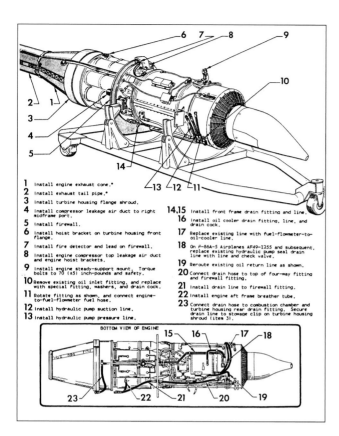

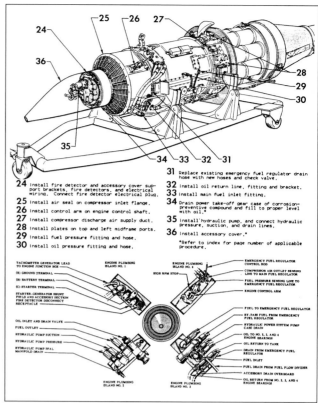

Left diagram parts list:

1. Install engine exhaust cone.*
2. Install exhaust tail pipe.*
3. Install turbine housing flange shroud.
4. Install compressor leakage air duct to right midframe port.
5. Install firewall.
6. Install hoist bracket on turbine housing front flange.
7. Install fire detector and lead on firewall.
8. Install engine compressor top leakage air duct and engine hoist brackets.
9. Install engine steady-support mount. Torque bolts to 70 (±5) inch-pounds and safety.
10. Remove existing oil inlet fitting, and replace with special fitting, washers, and drain cock.
11. Rotate fitting as shown, and connect engine-to-fuel-flowmeter fuel hose.
12. Install hydraulic pump suction line.
13. Install hydraulic pump pressure line.

14,15. Install front frame drain fitting and line.
16. Install oil cooler drain fitting, line, and drain cock.
17. Replace existing line with fuel-flowmeter-to-oil-cooler line.
18. On F-86A-5 Airplanes AF49-1255 and subsequent, replace existing hydraulic pump seal drain line with line and check valve.
19. Reroute existing oil return line as shown.
20. Connect drain hose to top of four-way fitting and firewall fitting.
21. Install drain line to firewall fitting.
22. Install engine aft frame breather tube.
23. Connect drain hose to combustion chamber and turbine housing rear drain fitting. Secure drain line to stowage clip on turbine housing shroud (item 31).

BOTTOM VIEW OF ENGINE

Right diagram parts list:

24. Install fire detector and accessory cover support brackets, fire detectors, and electrical wiring. Connect fire detector electrical plug.
25. Install air seal on compressor inlet flange.
26. Install control arm on engine control shaft.
27. Install compressor discharge air supply duct.
28. Install plates on top and left midframe ports.
29. Install fuel pressure fitting and hose.
30. Install oil pressure fitting and hose.

31. Replace existing emergency fuel regulator drain hose with new hoses and check valve.
32. Install oil return line, fitting and bracket.
33. Install main fuel inlet fitting.
34. Drain power take-off gear case of corrosion-preventive compound and fill to proper level with oil.*
35. Install hydraulic pump, and connect hydraulic pressure, suction, and drain lines.
36. Install accessory cover.*

*Refer to index for page number of applicable procedure.

TACHOMETER GENERATOR LEAD TO ENGINE JUNCTION BOX
(B) GROUND TERMINAL
(A) BATTERY TERMINAL
(C) STARTER TERMINAL
STARTER-GENERATOR SHUNT FIELD AND ACCESSORY SECTION FIRE DETECTOR DISCONNECT RECEPTACLE

ENGINE PLUMBING ISLAND NO. 1
ENGINE PLUMBING ISLAND NO. 4
HIGH RPM STOP

EMERGENCY FUEL REGULATOR CONTROL ROD
COMPRESSOR AIR OUTLET SENSING LINE TO MAIN FUEL REGULATOR
FUEL PRESSURE SENSING LINE TO EMERGENCY FUEL REGULATOR
ENGINE CONTROL ARM

OIL INLET AND DRAIN VALVE
FUEL OUTLET
HYDRAULIC PUMP SUCTION
HYDRAULIC PUMP PRESSURE
HYDRAULIC PUMP SEAL MANIFOLD DRAIN

FUEL TO EMERGENCY FUEL REGULATOR
BY-PASS FUEL FROM EMERGENCY FUEL REGULATOR
HYDRAULIC POWER SYSTEM PUMP CASE DRAIN
OIL TO NO. 1, 2, 3 AND 4 ENGINE BEARINGS
OIL RETURN TO TANK
DRAIN FROM EMERGENCY FUEL REGULATOR
FUEL INLET
FUEL DRAIN FROM FUEL FLOW DIVIDER
ACCESSORY DRAIN OVERBOARD
OIL RETURN FROM NO. 1, 2, AND 4 ENGINE BEARINGS

ENGINE PLUMBING ISLAND NO. 2
ENGINE PLUMBING ISLAND NO. 3

ABOVE LEFT AND RIGHT These engine build-up instructions list and illustrate the components that make up an engine change unit.

BELOW LEFT AND RIGHT The J47 is a reciprocating turbo-jet engine. Fuel is sprayed into the combustion cans where it is ignited by 'spark plugs' (self-sustaining after light-up). The fuel/air mix then expands rapidly thereby driving the aircraft forward. The expanding hot gases also drive the turbine, which in turn drives the compressor section, which sucks in air and compresses it for the combustion chamber where it is mixed with fuel and thus the process repeats itself. *(Author)*

rotating and reciprocating parts moving under great stress, the jet engine has essentially only one moving part, namely the compressor/turbine shaft. This component spins smoothly on bearings under comparatively low stress. Consequently, and principally because of its simplicity (there was not much to go wrong), reliability of the engine was very high. Overhaul life (assessed on condition) for the J47 ranged from 15 hours (in 1948) to a theoretical 1,200 hours (625 achievable in practice) in 1956. For example, the J47-GE-23 was rated to run

ABOVE Engine rpm and exhaust-gas temperature indicators are probably the most important of all on the instrument panel. The engine's exhaust temperature has to be controlled entirely by the pilot.

LEFT It is vitally important to protect the engine from ingesting any foreign objects. The fitting of an intake blank is just one measure taken to prevent catastrophic damage caused by stones or other 'litter' being thrown up by other aircraft or the wind. The intake is carefully checked for foreign objects during the walk-around inspection. *(Author)*

225 hours between overhauls. As installed on the F-86F, it experienced one in-flight shut-down every 33,000 hours in 1955 and 1956. Compared to failure rates associated with piston engines this was truly impressive.

If handled correctly, engine mechanical failures were very rare: most 'engine failures' were a result of running out of fuel or were down to the pilot not respecting the handling limits (particularly in terms of engine acceleration). Pilot-induced engine surges or poorly executed starts could easily melt the turbine blades or destroy the compressor stages.

Another possible cause of engine failure could be as a result of foreign objects entering the engine. Birds, stones or any other substance

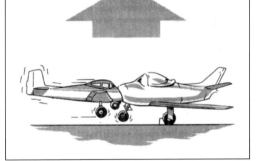

WARNING

Before starting engine, make sure danger areas fore and aft of airplane are clear of personnel, aircraft, and vehicles. (See figure 2-3.) Suction at the intake duct is of sufficient magnitude to kill or seriously injure personnel if they are drawn into or pulled suddenly against the duct. Danger aft of the airplane is created by the high exhaust temperature and blast from the tail pipe.

CAUTION

Engine thrust output decreases in direct pro-portion to tail-pipe temperature when oper-ating at constant rpm. For every 25-degree centigrade drop in tail-pipe temperature, there is a corresponding loss of approximately 200 pounds of thrust at sea level.

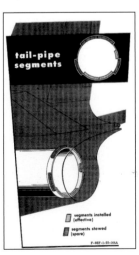

LEFT AND ABOVE Metal segments, commonly known as 'rats' or 'mice', could be inserted into the tailpipe opening to reduce or increase the area and thereby fine tune the exhaust temperature. Setting the temperature as close to the limit as possible would increase thrust and consequently add a few miles per hour to the top speed. *(Author)*

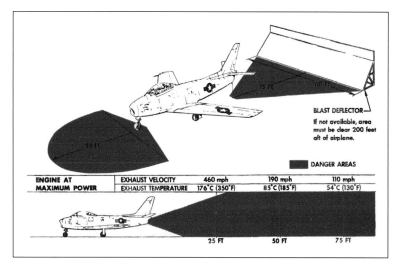

ENGINE AT MAXIMUM POWER	EXHAUST VELOCITY	460 mph	190 mph	110 mph
	EXHAUST TEMPERATURE	176°C (350°F)	85°C (185°F)	54°C (130°F)

ABOVE AND RIGHT Jet operations presented new threats and dangers to those accustomed to propeller-driven aircraft operations. The Sabre could cause serious injury or even death from either end.

(Author)

entering the intake would go right through the core of the engine usually causing a catastrophic failure, thus the J47 was highly susceptible to damage, especially when operating from unswept or semi-prepared surfaces. In contrast, modern turbo-fan engines allow some ingested debris to pass through the compressor bypass thereby minimising the damage to critical parts.

Oil system

Lubrication is provided by a pressure-type oil system with scavenge pump returning oil to a 5.7 US gallon tank. The only indication to the

LEFT The J47 oil system is self-contained on the engine and consists of an oil tank that feeds oil by gravity to a high-pressure pump that then supplies oil to the engine accessories gearbox and four engine shaft bearings. Scavenge pumps return the oil to the tank via a fuel-cooled oil cooler.

BELOW The oil pump output is directly related to engine rpm. At idle the pressure can be as low as 3 or 4psi, rising to about 40psi at full power. *(Author)*

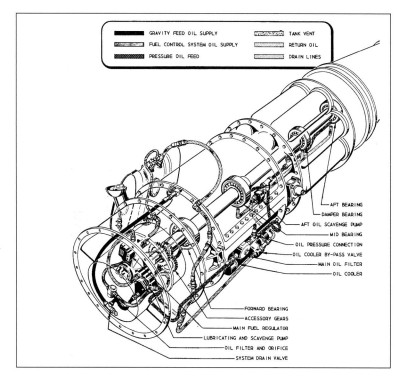

GRAVITY FEED OIL SUPPLY	TANK VENT
FUEL CONTROL SYSTEM OIL SUPPLY	RETURN OIL
PRESSURE OIL FEED	DRAIN LINES

AFT BEARING
DAMPER BEARING
AFT OIL SCAVENGE PUMP
MID BEARING
OIL PRESSURE CONNECTION
OIL COOLER BY-PASS VALVE
MAIN OIL FILTER
OIL COOLER

FORWARD BEARING
ACCESSORY GEARS
MAIN FUEL REGULATOR
LUBRICATING AND SCAVENGE PUMP
OIL FILTER AND ORIFICE
SYSTEM DRAIN VALVE

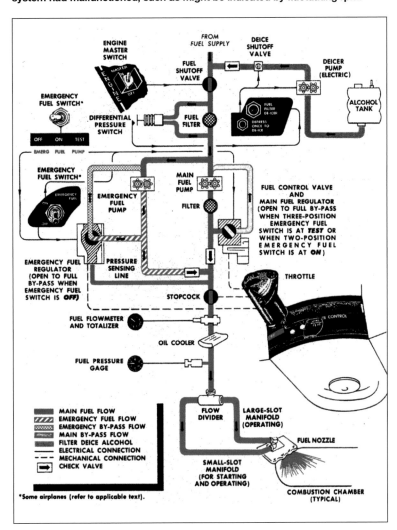

ABOVE The F-86A's oil tank is bolted to the airframe and is accessible by removing the inspection cover on the right-hand side of the fuselage above the wing. *(Author)*

LEFT At idle power the main fuel system is running at a pressure of about 40psi; at full power this can rise to over 400psi. *(Author)*

BELOW The main and emergency fuel systems work in parallel, the main system running at a slightly higher pressure. Any pressure drop in the main system would therefore allow the emergency system to take over control. A switch enabled the pilot to force this changeover if he suspected that the main system had malfunctioned, such as might be indicated by fluctuating rpm.

pilot is via a mechanical oil-pressure gauge. No manual control of the system is provided. The oil type (Aero Shell Turbine Oil 2) is very light and has a low viscosity.

Main engine control system

The main engine control system comprises an engine-driven positive-displacement fuel pump, a fuel control valve and a fuel flow regulator. The fuel regulator controls the amount of fuel that is delivered to the combustion chambers by operating the fuel control valve, which bypasses any fuel not required by the engine. Fuel requirements are determined by throttle position, engine speed, altitude and airspeed. Any throttle position above 'idle' sets a wide-range rpm governor, which operates through the regulator to maintain the selected rpm and provide overspeed protection. This regulator automatically compensates for changes in altitude and airspeed to maintain a constant rpm. It is a characteristic of jet engines that the possibility of flame-out increases with altitude. To reduce this risk the fuel regulator is designed to hold a fixed minimum fuel pressure sufficient to maintain combustion at all altitudes – in effect the minimum rpm increases at altitude. Above 40,000ft the engine

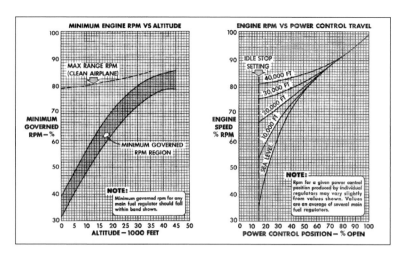

ABOVE J47-GE-7 fuel-regulator characteristics. The left-hand graph shows that the minimum governed rpm (to prevent flame-out) increases with altitude. The right-hand graph shows that at 40,000ft even with the throttle at idle the fuel regulator will maintain a rpm of not below about 83%. The fuel-regulator function of maintaining minimum rpm with change in altitude is just about the only automatic feature controlling the engine.

in airspeed, temperature, density of engine inlet air and changing thermal conditions within the engine. Generally, an increase in outside air temperature or altitude tends to cause gas exhaust temperature to rise, whereas increased airspeed causes it to fall. All three factors can change simultaneously and the effect on exhaust temperatures can be inconsistent. At about 35,000ft, exhaust temperatures will start to rise and continue to increase up to 45,000ft where they might exceed the limit at 100% rpm. The pilot, therefore, has to keep an eye on the temperature and retard the throttle to keep within limits.

With the main fuel control system in use, overspeed and flame-out protection (minimum rpm) are the only automatic assistances afforded the pilot: the job of keeping the engine working within temperature limits and managing surge-free accelerations is all down to the pilot's left hand.

Emergency engine control system

The emergency fuel system provides altitude compensation to maintain constant rpm with a change in altitude; however, it doesn't provide any rpm governing, temperature limiting or flame-out or surge protection. It is a very basic engine control system intended only to provide enough control to the pilot to 'get him

is only controllable in the top quadrant of the throttle with a high rpm maintained even with the throttle in the 'idle' position. At such high altitudes the throttle must be moved very carefully to avoid flame-out.

Another problem that manifests itself at high altitude is the characteristic for jet engines with fixed area exhaust nozzles to vary their exhaust temperatures dependent on changes

RIGHT AND FAR RIGHT If the main fuel controller fails and the emergency system is also defective then a dangerous overspeed condition can occur. For this reason a test of the emergency fuel system is carried out on the ground before take-off. The emergency fuel pump runs at a lower pressure than the main system, thus the maximum rpm achievable in emergency operation will be less and must be within the tolerance shown on this chart.

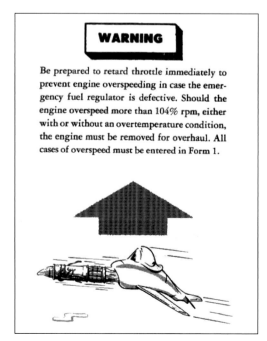

WARNING

Be prepared to retard throttle immediately to prevent engine overspeeding in case the emergency fuel regulator is defective. Should the engine overspeed more than 104% rpm, either with or without an overtemperature condition, the engine must be removed for overhaul. All cases of overspeed must be entered in Form 1.

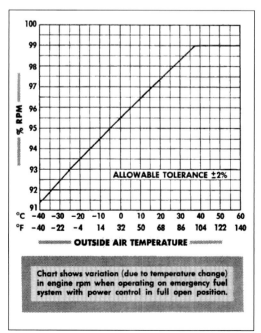

Chart shows variation (due to temperature change) in engine rpm when operating on emergency fuel system with power control in full open position.

ABOVE The emergency fuel-control switch is located on the upper left side of the instrument panel. *(Author)*

ABOVE The emergency regulator (item number 8 in the diagram of the fuel-regulation and injection system above right) from the J47 is designed to provide the pilot with a 'get-you-home' capability and as such has very limited functionality. *(Author)*

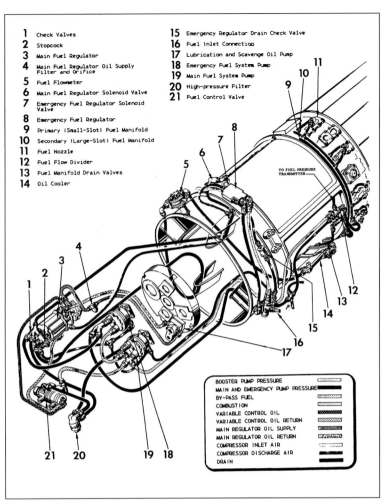

1	Check Valves	15	Emergency Regulator Drain Check Valve
2	Stopcock	16	Fuel Inlet Connection
3	Main Fuel Regulator	17	Lubrication and Scavenge Oil Pump
4	Main Fuel Regulator Oil Supply Filter and Orifice	18	Emergency Fuel System Pump
5	Fuel Flowmeter	19	Main Fuel System Pump
6	Main Fuel Regulator Solenoid Valve	20	High-pressure Filter
7	Emergency Fuel Regulator Solenoid Valve	21	Fuel Control Valve
8	Emergency Fuel Regulator		
9	Primary (Small-Slot) Fuel Manifold		
10	Secondary (Large-Slot) Fuel Manifold		
11	Fuel Nozzle		
12	Fuel Flow Divider		
13	Fuel Manifold Drain Valves		
14	Oil Cooler		

TO FUEL PRESSURE TRANSMITTER

BOOSTER PUMP PRESSURE	
MAIN AND EMERGENCY PUMP PRESSURE	
BY-PASS FUEL	
COMBUSTION	
VARIABLE CONTROL OIL	
VARIABLE CONTROL OIL RETURN	
MAIN REGULATOR OIL SUPPLY	
MAIN REGULATOR OIL RETURN	
COMPRESSOR INLET AIR	
COMPRESSOR DISCHARGE AIR	
DRAIN	

ABOVE Detailed view of the fuel-regulation and injection system.

home' in the event that the main system fails. All parameters and limits are therefore controlled completely manually by the pilot through the throttle. The system is set at a lower pressure than the main system and therefore under normal operations is bypassed. However, if main system pressure fails or drops then the emergency regulator provides control of the engine. The system can be tested before take-off or manually activated in the air by a switch located on the upper left coaming.

The power control

The power control (or throttle lever) is located on the left of the cockpit and falls nicely to hand. As well as serving the function of adjusting the engine speed, it also provides a way to completely cut off the fuel supply to

LEFT The power control lever, otherwise known to the pilots as the throttle. This lever is mechanically connected to both the main and the emergency fuel regulators. Additionally, it controls micro switches that command the operation of the fuel booster pumps and energise the ignition relays. This control is always moved with great caution. *(Author)*

RIGHT All the controls required for engine starting and relighting are located on the right-hand console.
(Author)

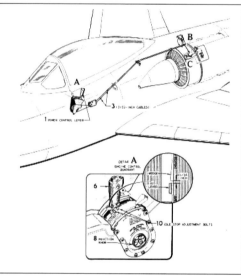

RIGHT AND BELOW Control of the engine is mechanical by cables and push rods connecting the power control level to the fuel-control valve. No electrical power whatsoever is required for engine control.
(Author)

the engine through a 'gate' that can be passed by rocking the lever outboard. The grip on the power control contains an A-1 CM gun sight electrical caging button, a microphone button and a speed-brake switch. Furthermore, on aircraft with the MK 18 gun sight, rotation of the power control grip operates manual ranging control of the sight.

Completing the engine controls are: the engine master switch (which electrically energises the booster pumps, operates a fuel shut-off valve and powers the ignition circuits); the emergency ignition switch for relighting in the air; and the starter generator control switch. The latter component behaves as a motor to drive the engine during start, then at approximately 23% rpm it switches to become a generator producing 28V DC.

Fuel system

There are five self-sealing fuel tanks installed in the F-86A: two in the fuselage, one in the wing centre section and one in each outer wing panel. Fuel is supplied under pressure from the centre wing tank to the engine by means of two electrical booster pumps. The centre wing tank receives fuel from all other internal tanks by gravity feed. Gravity flow from the rear fuselage tanks is supplemented by a transfer pump that is automatically actuated when the fuel level in the centre wing tank drops below approximately 56 gallons. The main fuel supply can be augmented by fitting a 120-gallon (or 206-gallon) drop tank under each wing. These drop tanks can be jettisoned if necessary: indeed, this was standard procedure before entering air-to-air combat. Fuel from the drop tanks is routed to the forward fuselage tank (through a fuel-level control valve) when the fuel level in the fuselage tank decreases by approximately 5 gallons. The transfer of fuel from the drop tanks is facilitated by air pressure bled from the engine compressor. It is important to remember to isolate this pressure line by switching off the pressure shut-off valve (located in the left-hand cockpit console) when on the ground to prevent fuel expanding in the drop tanks and then back-flowing into the engine compressor.

Alcohol from a small tank can provide de-icing of a frozen fuel filter by injecting

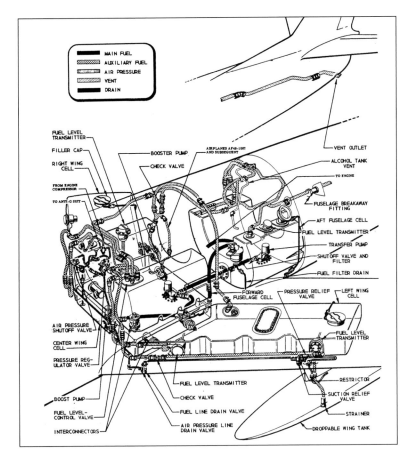

Legend:
- MAIN FUEL
- AUXILIARY FUEL
- AIR PRESSURE
- VENT
- DRAIN

FUEL LEVEL TRANSMITTER
FILLER CAP
RIGHT WING CELL
FROM ENGINE COMPRESSOR
TO ANTI-G SUIT
AIR PRESSURE SHUTOFF VALVE
CENTER WING CELL
PRESSURE REGULATOR VALVE
BOOST PUMP
FUEL LEVEL-CONTROL VALVE
INTERCONNECTORS

BOOSTER PUMP
CHECK VALVE
AIRPLANES AF49-1007 AND SUBSEQUENT
FORWARD FUSELAGE CELL
FUEL LEVEL TRANSMITTER
CHECK VALVE
FUEL LINE DRAIN VALVE
AIR PRESSURE LINE DRAIN VALVE

VENT OUTLET
ALCOHOL TANK VENT
TO ENGINE
FUSELAGE BREAKAWAY FITTING
AFT FUSELAGE CELL
FUEL LEVEL TRANSMITTER
TRANSFER PUMP
SHUTOFF VALVE AND FILTER
FUEL FILTER DRAIN
PRESSURE RELIEF VALVE
LEFT WING CELL
FUEL LEVEL TRANSMITTER
RESTRICTOR
SUCTION RELIEF VALVE
STRAINER
DROPPABLE WING TANK

ABOVE Internal view of the fuel system showing the location of the various fuel cells, pumps and valves. All fuel from the droppable wing tanks and integral fuel cells is eventually fed (by gravity, engine bleed air pressure, or electric pumps) to the centre wing cell from where it is picked up and pumped to the main and emergency engine-fuel control systems.

BELOW Fuel quantity is displayed in US gallons and only internal fuel is indicated. *(Author)*

BELOW The F-86A had an internal 'useable' fuel capacity of 435 US gallons.

FUEL QUANTITY DATA U.S. GALLONS

TANK	NO.	USABLE FUEL	EXPANSION SPACE	UNUSABLE FUEL	TOTAL VOLUME
FORWARD FUSELAGE AND CENTER WING TANK	1	196	17	5	218
AFT FUSELAGE TANK	1	105	0	1	106
LEFT WING TANK	1	67	0	1	68
RIGHT WING TANK	1	67	0	1	68
TOTAL INTERNAL FUEL		435	17	8	460
DROP TANKS LARGE SMALL	2 2	206 120	*43 4	3 0	252 124

*Tank capacity is 252 gallons, although tanks can be filled only to 206½ gallon level.

BASED ON TEST DATA

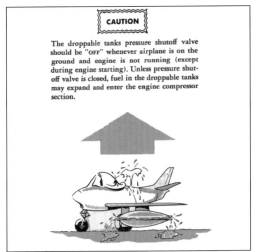

ABOVE AND BELOW If pilots or ground crew forgot to switch off the drop-tank pressure shut-off switch, fuel in the drop tanks could expand (especially on a warm day) and make its way into the engine compressor. This would make the next start attempt hazardous and somewhat pungent.

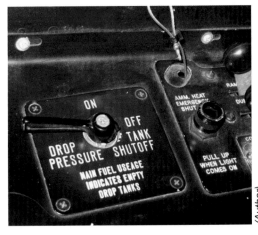

(Author)

Warning

IF PRESSURE SHUTOFF VALVE IS *ON* WHEN AIRPLANE IS ON GROUND AND ENGINE ISN'T RUNNING, DROP TANK FUEL MAY EXPAND AND ENTER COMPRESSOR SECTION, CREATING A POTENTIAL EXPLOSION HAZARD.

ABOVE One of the Sabre's 120-gallon drop tanks. It was always envisaged that external fuel tankage would be needed; however, the change from straight to swept-wings made it difficult to mount such tanks on the wing-tips (as was the fashion) and still maintain a satisfactory centre of gravity. *(Author)*

ABOVE External fuel tanks were pylon-mounted on the wing as far inboard as practically possible to minimise any change in centre of gravity as the fuel was used up. *(John Dibbs)*

ABOVE The F-86 without drop tanks: a rare configuration. The Sabre is usually operated with external tanks fitted (2 x 120 US gallons) to provide greater range and operational flexibility. *(Colin Norwood)*

Warning

KEEP CLEAR OF TAIL PIPE AND DO NOT MOVE AIRPLANE INTO HANGAR FOR AT LEAST 15 MINUTES AFTER SHUTDOWN, BECAUSE OF THE POSSIBILITY OF EXPLOSION DUE TO THE ACCUMULATION OF FUEL VAPORS.

BELOW When refuelling the aircraft the forward fuselage tank must be filled first in order to utilise the full capacity of the fuel system. *(Author)*

FUEL FILLER 195 US GAL

FILL FWD. FUS. TANK FIRST CAUTION - DO NOT OVERFLOW

LEFT Intake and tailpipe blanks are usually not fitted for about 15 minutes after shut-down to allow fuel vapours to escape into the atmosphere.

BELOW Fuel system schematic.

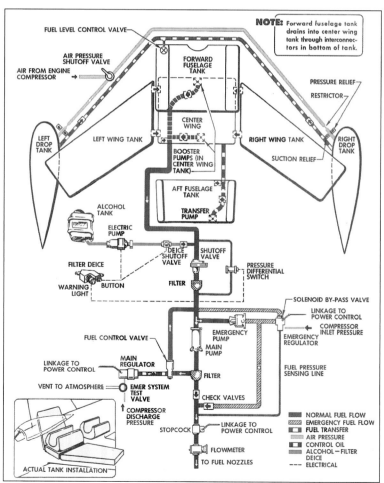

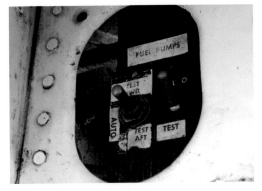

alcohol into the fuel entering the filter for up to 1 minute, hopefully enough time for the pilot to descend to a warmer level. Fuel-filter icing is shown by an amber warning light on the left coaming and the fluid is pumped by an electrical motor activated by the pilot pushing and holding a button until the ice warning extinguishes.

Electrical system

Electrical energy is supplied by a 28V direct-current system, powered by a 400A engine-driven starter generator. A 24V, 34A hour battery serves as a standby, supplying power to part of the system when the generator is inoperative or generator output is insufficient to close the reverse-current relay. Alternating current is supplied by a three-phase inverter, a single-phase inverter and a three-phase electronic inverter of the A-1 CM gun sight.

ABOVE AND LEFT The generator field control relay and voltage regulator are located behind an inspection panel on the left-hand side of the fuselage below the cockpit sill. *(Author)*

ABOVE When restored, 48-178 was fitted with an improved electrical arrangement and battery-charging system installed in the left-hand gun bay. This allows the aircraft to carry out internal battery starts, a frequent requirement at some minor airfields. (Author)

ABOVE RIGHT Electrical controls are located on the forward right-hand console and comprise the battery switch, generator on/off/reset, voltage regulator rheostat (hidden behind the 'warning' flap), battery start mode switch (start or charge) and the battery-charge status indicator lights. (Author)

RIGHT The generator output voltage can be adjusted by the rheostat on the forward right-hand console to give 28V. A marker needle on the gauge indicates the required voltage. (Author)

ABOVE Over-voltage, generator failure, generator overload, battery mode (start or charge) and inverter failure are indicted by warning lights and an ammeter on the lower left of the instrument panel. (Author)

BELOW The main circuit-breaker panel is located on the cockpit wall next to the pilot's left elbow (the anti-'g' suit valve test button is on the left of the picture). A smaller breaker panel is located on the other side of the cockpit next to the pilot's right elbow.

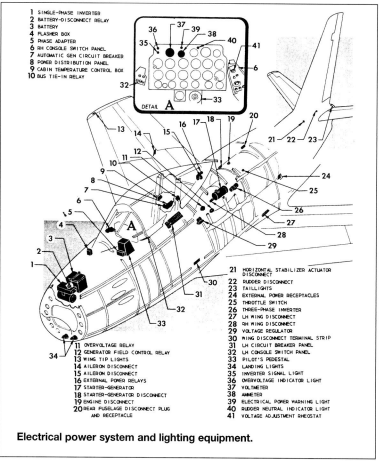

1 SINGLE-PHASE INVERTER
2 BATTERY-DISCONNECT RELAY
3 BATTERY
4 FLASHER BOX
5 PHASE ADAPTER
6 RH CONSOLE SWITCH PANEL
7 AUTOMATIC GEN CIRCUIT BREAKER
8 POWER DISTRIBUTION PANEL
9 CABIN TEMPERATURE CONTROL BOX
10 BUS TIE-IN RELAY

11 OVERVOLTAGE RELAY
12 GENERATOR FIELD CONTROL RELAY
13 WING TIP LIGHTS
14 AILERON DISCONNECT
15 AILERON DISCONNECT
16 EXTERNAL POWER RELAYS
17 STARTER-GENERATOR
18 STARTER-GENERATOR DISCONNECT
19 ENGINE DISCONNECT
20 REAR FUSELAGE DISCONNECT PLUG
    AND RECEPTACLE

21 HORIZONTAL STABILIZER ACTUATOR DISCONNECT
22 RUDDER DISCONNECT
23 TAILLIGHTS
24 EXTERNAL POWER RECEPTACLES
25 THROTTLE SWITCH
26 THREE-PHASE INVERTER
27 LH WING DISCONNECT
28 RH WING DISCONNECT
29 VOLTAGE REGULATOR
30 WING DISCONNECT TERMINAL STRIP
31 LH CIRCUIT BREAKER PANEL
32 LH CONSOLE SWITCH PANEL
33 PILOT'S PEDESTAL
34 LANDING LIGHTS
35 INVERTER SIGNAL LIGHT
36 OVERVOLTAGE INDICATOR LIGHT
37 VOLTMETER
38 AMMETER
39 ELECTRICAL POWER WARNING LIGHT
40 RUDDER NEUTRAL INDICATOR LIGHT
41 VOLTAGE ADJUSTMENT RHEOSTAT

Electrical power system and lighting equipment.

LEFT The battery bay in the nose of the Sabre is also the home of a single-phase inverter and battery disconnect relay. *(Author)*

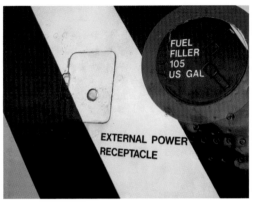

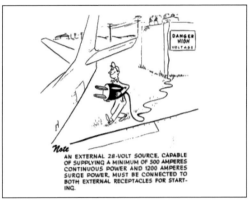

FAR LEFT AND LEFT An external power receptacle is located on the left side of the fuselage above the trailing edge of the wing. *(Author)*

BELOW Electrical system schematic.

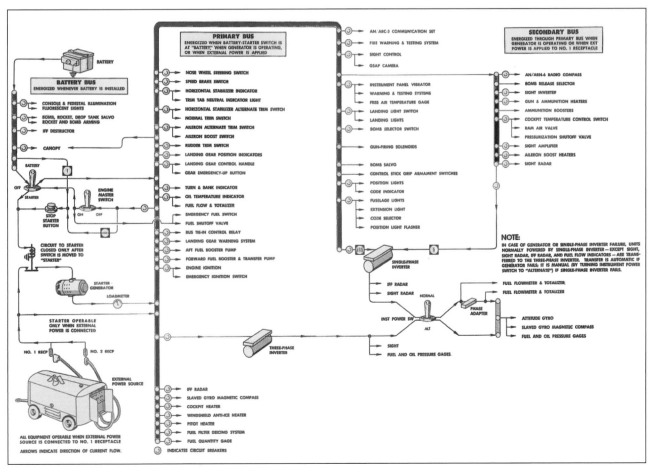

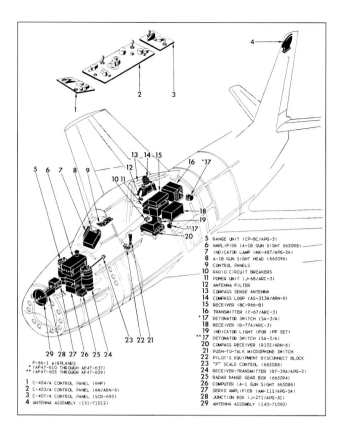

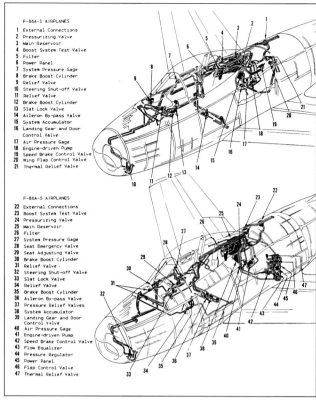

F-86-1 AIRPLANES

5 RANGE UNIT (CP-8C/APG-5)
6 AMPLIFIER (A-1B GUN SIGHT 665098)
7 INDICATOR LAMP (MX-487/APG-5A)
8 A-1B GUN SIGHT HEAD (665096)
9 CONTROL PANELS
10 RADIO CIRCUIT BREAKERS
11 POWER UNIT (J-68/ARC-3)
12 ANTENNA FILTER
13 COMPASS SENSE ANTENNA
14 COMPASS LOOP (AS-313A/ARN-6)
15 RECEIVER (BC-966-B)
16 TRANSMITTER (T-67/ARC-3)
*17 DETONATOR SWITCH (SA-3/A)
18 RECEIVER (R-77A/ARC-3)
19 INDICATOR LIGHT (FOR IFF SET)
**17 DETONATOR SWITCH (SA-3/A)
20 COMPASS RECEIVER (R101/ARN-6)
21 PUSH-TO-TALK MICROPHONE SWITCH
22 PILOT'S EQUIPMENT DISCONNECT BLOCK
23 "F" SCALE CONTROL (665088)
24 RECEIVER-TRANSMITTER (RT-39A/APG-5)
25 RADAR RANGE GEAR BOX (665094)
26 COMPUTER (A-1 GUN SIGHT 665086)
27 SERVO AMPLIFIER (AM-111/APG-5A)
28 JUNCTION BOX (J-271/APG-5C)
29 ANTENNA ASSEMBLY (140-71090)

F-86-1 AIRPLANES
* (AF47-610 THROUGH AF47-637)
** (AF47-605 THROUGH AF47-609)

1 C-404/A CONTROL PANEL (VHF)
2 C-403/A CONTROL PANEL (AN/ARN-6)
3 C-407/A CONTROL PANEL (SCR-695)
4 ANTENNA ASSEMBLY (151-71012)

F-86A-1 AIRPLANES

1 External Connections
2 Pressurizing Valve
3 Main Reservoir
4 Boost System Test Valve
5 Filter
6 Power Panel
7 System Pressure Gage
8 Brake Boost Cylinder
9 Relief Valve
10 Steering Shut-off Valve
11 Relief Valve
12 Brake Boost Cylinder
13 Slat Lock Valve
14 Aileron By-pass Valve
15 System Accumulator
16 Landing Gear and Door Control Valve
17 Air Pressure Gage
18 Engine-driven Pump
19 Speed Brake Control Valve
20 Wing Flap Control Valve
21 Thermal Relief Valve

F-86A-5 AIRPLANES

22 External Connections
23 Boost System Test Valve
24 Pressurizing Valve
25 Main Reservoir
26 Filter
27 System Pressure Gage
28 Seat Emergency Valve
29 Seat Adjusting Valve
30 Brake Boost Cylinder
31 Relief Valve
32 Steering Shut-off Valve
33 Slat Lock Valve
34 Relief Valve
35 Brake Boost Cylinder
36 Aileron By-pass Valve
37 Pressure Relief Valves
38 System Accumulator
39 Landing Gear and Door Control Valve
40 Air Pressure Gage
41 Engine-driven Pump
42 Speed Brake Control Valve
43 Flow Equalizer
44 Pressure Regulator
45 Power Panel
46 Flap Control Valve
47 Thermal Relief Valve

ABOVE LEFT The 'A' model Sabre's electronic systems were limited to the communications and weapons systems.

ABOVE RIGHT Hydraulic utility power and open-centre system for F-86A-1 (top) and F-86A-5 (bottom) aeroplanes.

Hydraulic system

The main hydraulic system combines a pressure-type (utility) system and an open-centre (boost) system. Output of a single engine-driven hydraulic pump is divided equally between the two systems by a flow divider.

The pressure-type system is under constant pressure and incorporates an accumulator for pressure storage and is controlled by a pressure regulating valve. The system powers landing gear, speed-brakes, nose-wheel steering and wheel-brake control boost. Early Sabres also used hydraulics for the wing-flap operation (later

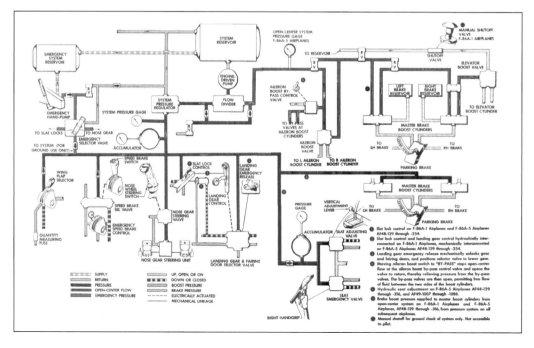

RIGHT Hydraulic system schematic.

ABOVE The main hydraulic system pressure gauge is located on the bottom of the instrument panel. This small instrument may look insignificant, but it is very important as a way of relaying to the pilot the work that the hydraulic system is doing. *(Author)*

FAR LEFT The hydraulic reservoir filler cap is just visible on the left of this inspection hatch.

LEFT The hydraulic reservoir is located on the right-hand side of the fuselage. The brass hexagonal filler cap and sight-level glass are visible behind the grey box. This box contains the vacuum tubes and circuitry that are part of the engine bay/tail-cone fire-detection system. *(Author)*

Flight control system

In the F-86A the primary flight controls are conventionally operated via push rods and control wires with a hydraulic boost system to reduce the amount of stick force required to move the ailerons and elevators.

LEFT The Sabre sports a comfortable multi-function control column that in addition to being the primary means of aircraft control carries switches for aircraft trimming, weapon release, gun firing, gun-sight caging and nose-wheel steering activation. *(Author)*

'A's and other variants used electric motors). The landing gear and speed-brake controls have an off position principally to prevent fluid loss in the event of a damaged hydraulic line.

The open-centre hydraulic system is a non-pressurised system with no means of pressure storage. When the engine-driven pump is running, fluid circulates through the system without building up pressure until a control valve is moved from neutral to an operating position. Then fluid circulation is restricted at the affected control valve to build up pressure required to perform the desired operation, namely surface control boost (aileron and elevator) and on early aircraft the wheel-brake control boost.

A pressure gauge for the pressure-type system is located on the main instrument panel, whereas the open-centre system and boost gauge are located on the right console under the pilot's right elbow.

ABOVE The hydraulic boost pressure gauge is located under the pilot's right elbow and is not easy to see. However, it only really needs to be checked on the ground during a functional check of the elevator and aileron boost system. (Author)

BELOW The aileron bypass switch is sited on the left-hand console behind the throttle quadrant. Primary yaw-trim and alternate roll-trim controls are also in this position. Alternate pitch trim is on the left-hand cockpit wall in front of the throttle. (Author)

RIGHT The horizontal stabiliser indicator is positioned just below the rudder trim neutral light at the top of the instrument panel. There is no indication in the cockpit of aileron-trim position and this is set to neutral by looking out at the left aileron where it can be seen that the trim tab is visibly lined up with the aileron and thereby neutral. (Author)

(On subsequent Sabres a much improved non-reversible dual hydraulic system and a new tail-plane design was employed which, although adding weight, improved the aircraft's handling markedly, especially at high speed.)

Should the boost system fail, stick forces are considerably increased. However, the aircraft remains flyable provided the ailerons are trimmed 'hands off' at the time of the failure.

In the event of a hydraulic failure the aileron boost system is automatically bypassed, or the system can also be manually switched off if desired. At very high speeds roll control without the boost system is almost impossible, although gentle roll rates can be achieved with trim and/or rudder. There is no bypass system

BELOW This pilot's eye view shows that the aileron trim is set to neutral. (Author)

for the elevator boost because this control has a much lower boost ratio than the aileron boost and the forces are easy to overcome by the pilot.

Rudder and aileron trim tabs are provided, but instead of conventional elevator trim the complete horizontal stabiliser is adjustable. Roll and pitch trimming is usually provided by the stick grip four-way switch. However, a backup system is also provided to cater for trim runaway conditions: these operate at half the speed of the main trimmers and are unsuitable for adjustment during manoeuvre, being intended primarily to provide an override to a runaway trim condition. During manoeuvring it is desirable to adjust the elevator trim, and the stick-top switch makes this easy and intuitive. The rudder is trimmed by a switch on the left console, for which there is no alternate system.

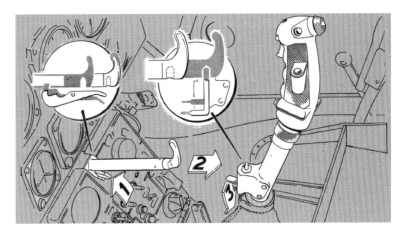

Early aircraft were equipped with a stall warning system (stick-shaker), but on later aircraft no stall warning system was necessary as sufficient natural stall warning was obtained due to a different wing-slat installation.

ABOVE A built-in surface-control lock for locking all the control surfaces connects to the control column and also locks the rudder pedals.

The '6-3' wing

Some Sabre variants dispensed with leading edge slats entirely, most notably late 'F' models that introduced the '6-3' wing, so-called because of extensions to the chord of 6in at the root and 3in at the tip. The introduction of this wing (fitted as a kit on Sabres in theatre), together with a mid-span aerodynamic fence mounted on the leading edge and a sharper leading edge profile, reduced weight, improved climb performance and provided better high-altitude turn performance. It also alleviated the unreliability problems that many pilots had experienced in combat in Korea whereby the slats would deploy when manoeuvring hard at high Mach number. These improvements, along with the 'F's more powerful engine, helped to redress the performance imbalance between the Sabre and MiG-15. The penalty was that stall characteristics were more vicious than those of the slatted wing (which is thought to have caused a number of accidents until better training was introduced), plus a slight increase in take-off and landing speed.

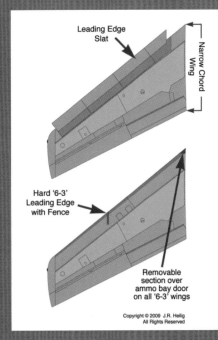

Leading Edge Slat

Narrow Chord Wing

Hard '6-3' Leading Edge with Fence

Removable section over ammo bay door on all '6-3' wings

Copyright © 2009 J.R. Heilig All Rights Reserved

LEFT The original Sabre wing design (top) was modified by the addition of a hard leading edge, wing fence and root/tip chord extension. This was fitted to some Sabres and (late F-86Fs) and licence-built aircraft (Canadair Mk 5 and CAC Avon Sabre). Other modifications included the reintroduction of the slats to the '6-3' wing (Canadair Mk 6) and a '6-3' wing with slats and a 12in longer wingspan (F-86F-40, H and L). *(J.R. Heilig 2009)*

LEFT The '6-3' wing modification extended the wing root by 6in and this then interfered with the opening of the ammunition bay door. As shown here, a small triangular section of the wing-root leading edge was removable to permit the door to be opened.

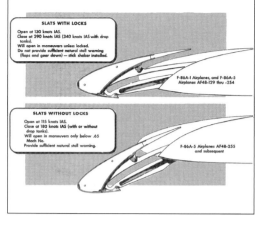

SLATS WITH LOCKS

Open at 130 knots IAS.
Close at 290 knots IAS (340 knots IAS with drop tanks).
Will open in maneuvers unless locked.
Do not provide sufficient natural stall warning (flaps and gear down) — stick shaker installed.

F-86A-1 Airplanes, and F-86A-5 Airplanes AF48-129 thru -254

SLATS WITHOUT LOCKS

Open at 115 knots IAS.
Close at 180 knots IAS (with or without drop tanks).
Will open in maneuvers only below .65 Mach No.
Provide sufficient natural stall warning.

F-86A-5 Airplanes AF48-255 and subsequent

ABOVE 48-178 is unusual and unique in that it has a standard 'A' wing, but with fixed leading edges (the slats have been disabled) and a wing fence (as was fitted to F-86Fs with the '6-3' wing). *(Author)*

ABOVE The design of the slats changed several times over the life of the F-86A due to unreliability and maintenance problems.

BELOW The fowler-type wing flaps are powerful and extend to about half the wingspan. *(John Myers, Author)*

Wing slats and flaps

Wing slats extend from the fuselage to wing-tip along the leading edge of each wing panel. Aerodynamic forces acting upon the slats cause them to open or close automatically depending on the airspeed and altitude of the aircraft. When they open, the slats move forward along a curved track forming a slot in the wing leading edge. This changes the airflow over the wing upper surface and increases lift resulting in a lower stall speed. At higher speed in unaccelerated flight the slats automatically close to provide minimum drag for maximum performance. Early aircraft had hydraulically activated locks to hold the slats in the closed (flush) position. Depending on the variant, the slat locks are linked either mechanically or hydraulically to the landing gear to ensure that the slots are unlocked with the gear in a symetric down position. This prevented deployment during manoeuvre.

The following refers to aircraft fitted with electrically powered wing flaps (F-86A-7 onwards).

Each wing flap is actuated by an individual electric circuit and electric motor. The flaps are mechanically interconnected so that if one system fails the flaps can still be moved by the opposite flap system, albeit at a reduced rate. The interconnection also prevents individual or uneven flap movement. Flap-up travel is limited by mechanical stops; flap-down travel is limited by electrical limit switches. A brake coil in each actuator prevents air loads from moving the flaps. No emergency system for flap operation is provided as there is adequate protection in the normal system through mechanical linkage, the individual actuators and electrical circuits.

The flap control is mounted inboard of the power control on the left console. To move the

FAR LEFT AND LEFT
Detail of the right-hand wing flap showing the flap guides tracks (left) and roller bearings that follow the tracks (right). *(Author)*

flaps to full-up or full-down position the flap lever is placed at 'UP' or 'DOWN'; it is not necessary to return to the hold position. For intermediate flap positions 'HOLD' is selected when the desired flap position is obtained. With the electrical flap system there is no flap position gauge; position can be seen by looking at either wing, or more usually it is 'felt' in the seat of the pants.

Speed-brakes

Hydraulically operated speed-brakes are located on each side of the fuselage below the dorsal fin. Each of the two brakes consists of a hinged panel which, when open, extends down and forward into the airstream.

The speed-brakes are controlled by a switch mounted on the top of the throttle, which controls the speed-brake hydraulic control valve. The switch has three positions: 'IN', 'OUT' and a neutral position. After the speed-brakes have been extended or retracted the switch should be returned to neutral to prevent loss of fluid in the event of a leak and, because the hydraulic lines are routed through the hot jet-pipe bay, to reduce any fire hazard.

LEFT Left-hand flap shown at full extension. *(Author)*

BELOW In this picture the Sabre's large speed-brakes, dihedral tail-plane, flaps and pop-down taxi/landing lights are all clearly shown. *(Neil Cotten)*

RIGHT Sabre on turn-a-round at the 2006 Koksijde air show in Belgium. Note the auxiliary undercarriage doors open and the external power supply connected. Flaps are retracted to improve access to the drop-tank filler caps for refuelling. The F-86 is endowed with large, powerful air-brakes; however, they can produce a strong nose-up trim change at high speed. *(Andrew Critchell, aviationphoto.co.uk)*

Since the hydraulic lines to the speed brake actuating cylinders are routed near the engine, it is extremely important that the speed brake switch be kept in the neutral position to minimize the fire hazard should a line be damaged.

ABOVE The speed-brakes can be closed (by air loads) in the event of a hydraulic failure. A wire-locked lever on the left console outboard of the power control dumps any hydraulic fluid, thereby removing any hydraulic lock. Once the speed-brakes have been retracted by use of the emergency system they cannot then be extended. *(Author)*

LEFT AND BELOW Detail of speed-brake panels, bays and hydraulic operating rams. *(Author)*

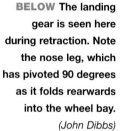

BELOW The landing gear is seen here during retraction. Note the nose leg, which has pivoted 90 degrees as it folds rearwards into the wheel bay. *(John Dibbs)*

Landing gear, wheel brakes and nose-wheel steering

The landing gear and wheel fairing doors are hydraulically actuated and electrically (DC) controlled and sequenced. On selection of 'UP' the doors open, the gear folds up and the doors close. The lowering sequence is the reverse. The main gear retracts inboard into the wing panels and fuselage; the nose gear retracts aft into the fuselage, pivoting to lie parallel with the bottom of the aeroplane.

Incorporated into the nose-gear assembly is a hydraulic steering unit, which acts as a conventional shimmy damper when the steering mechanism is not engaged.

A ground safety lock is provided for the nose-gear assembly to prevent inadvertent retraction

ABOVE The gear retraction sequence is captured partway through in this unusual shot. *(Walter van Bel)*

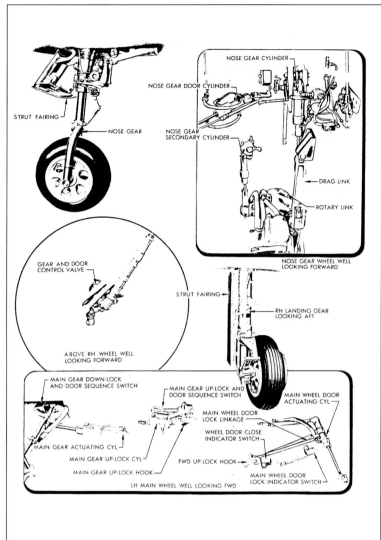

STRUT FAIRING — NOSE GEAR

NOSE GEAR CYLINDER

NOSE GEAR DOOR CYLINDER

NOSE GEAR SECONDARY CYLINDER

DRAG LINK

ROTARY LINK

NOSE GEAR WHEEL WELL LOOKING FORWARD

GEAR AND DOOR CONTROL VALVE

STRUT FAIRING

RH LANDING GEAR LOOKING AFT

ABOVE RH WHEEL WELL LOOKING FORWARD

MAIN GEAR DOWN-LOCK AND DOOR SEQUENCE SWITCH

MAIN GEAR UP-LOCK AND DOOR SEQUENCE SWITCH

MAIN WHEEL DOOR ACTUATING CYL

MAIN WHEEL DOOR LOCK LINKAGE

WHEEL DOOR CLOSE INDICATOR SWITCH

MAIN GEAR ACTUATING CYL

MAIN GEAR UP-LOCK CYL

FWD UP-LOCK HOOK

MAIN GEAR UP-LOCK HOOK

MAIN WHEEL DOOR LOCK INDICATOR SWITCH

LH MAIN WHEEL WELL LOOKING FWD

on the ground (something that the Sabre was sometimes minded to do) as there is no down lock as such. Instead, there is an over-centre scissor brace. No ground safety locks are provided for the main gear as the weight of the aircraft on the main gear down locks prevents accidental retraction when the aircraft is motionless.

Landing gear position lights are located on the left forward console. There are three green lights (one for each gear) and a single red light

GEAR AND WHEEL DOOR SOLENOID-OPERATED CONTROL VALVE

SHUTTLE VALVE

NOSE GEAR SECONDARY ACTUATING CYLINDER

NOSE GEAR ACTUATING CYLINDER

EMER PUMP HANDLE

MAIN GEAR UP-LOCK CYLINDER

MAIN GEAR DOWN-LOCK CYLINDER

MAIN WHEEL DOOR ACTUATING CYLINDER

LANDING GEAR CONTROL HANDLE

MAIN WHEEL DOORS

EMERGENCY GEAR LOWERING HANDLE

SHUTTLE VALVE

NOSE WHEEL DOOR

SHOCK STRUT FAIRING

NOSE WHEEL DOOR ACTUATING CYLINDER

SHOCK STRUT FAIRING

MAIN GEAR ACTUATING CYLINDER

ABOVE AND LEFT
The landing-gear system is highly reliant on several micro switches to correctly sequence the movement of the doors and gear legs. In an emergency the main gear can be extended by gravity and a blow-down bottle is provided to assist the forward lowering nose leg in overcoming air loads.

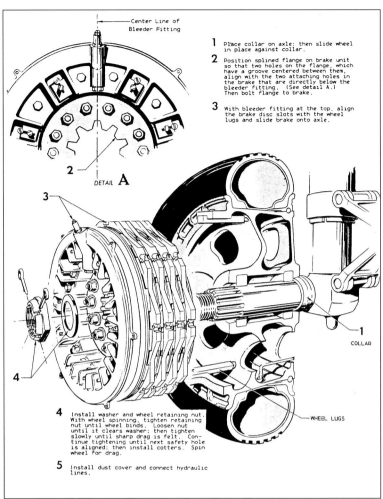

Center Line of Bleeder Fitting

1 Place collar on axle; then slide wheel in place against collar.

2 Position splined flange on brake unit so that two holes on the flange, which have a groove centered between them, align with the two attaching holes in the brake that are directly below the bleeder fitting. (See detail A.) Then bolt flange to brake.

3 With bleeder fitting at the top, align the brake disc slots with the wheel lugs and slide brake onto axle.

DETAIL **A**

COLLAR

WHEEL LUGS

4 Install washer and wheel retaining nut. With wheel spinning, tighten retaining nut until wheel binds. Loosen nut until it clears washer; then tighten slowly until sharp drag is felt. Continue tightening until next safety hole is aligned; then install cotters. Spin wheel for drag.

5 Install dust cover and connect hydraulic lines.

ABOVE Powerful multi-disc brakes are provided on the main wheels. These brakes do not have any anti-skid system and are modulated by the pilot using toe-operated motors mounted on the rudder pedals. *(Author)*

BELOW Specially commissioned for this book in the style of the original manufacturer's manuals, this illustrates the possible consequences of not fitting the ground safety brace to the nose landing-gear leg. *(First image credit Rob Gauld-Galliers, 2011)*

(connected to all three gear legs) to provide a visual indication of gear position. Each green light illuminates when its respective gear is down and locked. The red light illuminates when any gear is in any position other than down and locked or up and locked; or if the gear is up and locked and the throttle (power control) is retarded below minimum cruising rpm (when a warning horn also sounds); or if the gear is up and locked and any gear door is not completely closed. The warning

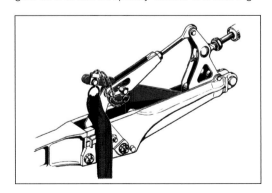

CAUTION

THE NOSE GEAR LOCKING PIN MUST BE FITTED BEFORE TOWING. THE NOSE LEG IS LIKELY TO COLLAPSE UNLESS SECURED.

LEFT The normal landing-gear controls are all grouped on the left console. The landing-gear handle electrically controls the landing gear and door hydraulic selector valve. The handle has three positions: 'UP', 'COMBAT' (neutral) and 'DOWN'. A ground safety switch prevents inadvertent retraction when the weight is on the wheels if the handle is moved to the 'UP' position. This can be intentionally overridden by pressing the emergency up button if the gear needs to be collapsed on the ground. Electrical and hydraulic power must be available for this function to work. The 'gear unsafe' light will illuminate and a warning horn will sound if the throttle is retarded below about 70% rpm. The amber light reminds the pilot to move the handle to 'COMBAT' after all the landing-gear doors are closed after retraction: if the handle is left in 'UP' then negative 'g'-loads may actuate the gear sequence switch and cause the gear doors to open and close which, at high speed, would damage the doors. *(Author)*

horn can be silenced by pressing a reset button (to reset the warning), by increasing the power above minimum cruise or by selecting the gear down and obtaining three greens.

If the landing gear fails to lower normally it may be lowered by use of the emergency release handle located at the bottom of the centre pedestal. When the handle is pulled to its full extension the main gear and all doors are mechanically unlocked. If the hydraulic system has failed, the main gear will fall free under gravity. However, the nose gear, although unlocked by the release handle, extends into the airflow and therefore can only achieve a

ABOVE Located at the bottom of the centre console is the emergency landing-gear extension handle. *(Author)*

LEFT A nitrogen bottle located in the nose undercarriage bay provides a 'kick' to the nose leg to extend it into the airflow in the event of a hydraulic failure. *(Author)*

ABOVE Detailed views of F-86A landing gear. *(All author)*

RIGHT The newly refurbished nose-wheel bay looking forward shows the nose-wheel steering cams and plumbing along with the up, down and auxiliary gear door hydraulic rams.

FAR RIGHT Right-hand main landing gear stripped for maintenance. The doors attached to the leg move purely mechanically.

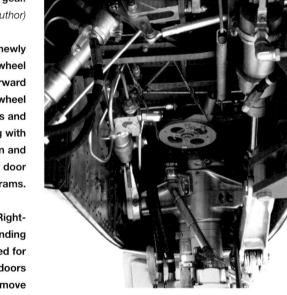

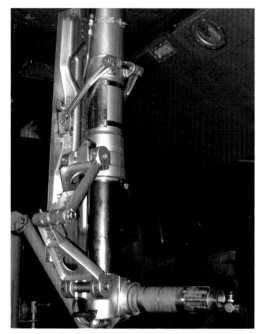

RIGHT AND FAR RIGHT Main and nose auxiliary landing-gear doors are usually opened after shut-down to allow ground crew to access the wheel bays for servicing purposes.

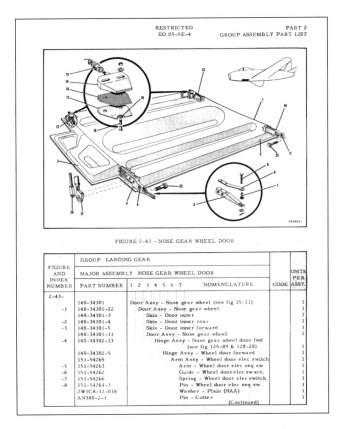

FIGURE 2-43 - NOSE GEAR WHEEL DOOR

FIGURE AND INDEX NUMBER	GROUP LANDING GEAR									UNITS PER ASSY.
	MAJOR ASSEMBLY NOSE GEAR WHEEL DOOR									
	PART NUMBER	1 2 3 4 5 6 7	NOMENCLATURE						CODE	
2-43-	140-34301		Door Assy - Nose gear wheel (see fig 35-11)							1
-1	140-34301-22		Door Assy - Nose gear wheel							1
	140-34301-3		Skin - Door outer							1
-2	140-34301-4		Skin - Door inner rear							1
-3	140-34301-5		Skin - Door inner forward							1
	140-34301-11		Door Assy - Nose gear wheel							1
-4	140-34302-13		Hinge Assy - Nose gear wheel door fwd (see fig 125-89 & 128-28)							1
	140-34302-5		Hinge Assy - Wheel door forward							1
	151-54265		Arm Assy - Wheel door elec switch							1
-5	151-54263		Arm - Wheel door elec seq sw							1
-6	151-54262		Guide - Wheel door elec sw act.							1
-7	151-54266		Spring - Wheel door elec switch							1
-8	151-54264-3		Pin - Wheel door elec seq sw							1
	2W1C8-12-018		Washer - Plain (NAA)							1
	AN380-2-1		Pin - Cotter							1
			(Continued)							

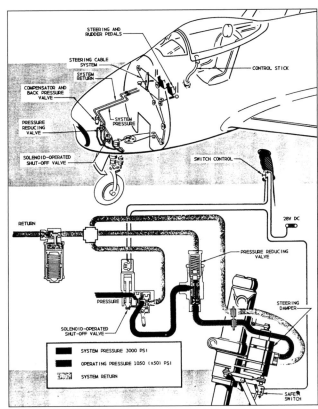

locked-down condition against these air loads by the assistance of gas pressure from a nitrogen bottle dedicated to this task.

Hydraulic pressure is supplied to a nose-gear steering unit through a shut-off valve controlled by a push button on the control stick grip. When the switch is depressed, with nose-wheel on the ground and the shock strut compressed, the nose-wheel can be turned approximately 21 degrees either side of centre by rudder pedal action. Before the mechanism can be engaged the rudder pedals must be coordinated with nose-wheel position.

Engine-fire detection system

Warning of an engine fire is given by indicator lights on the shroud above the right side of the instrument panel. Lights marked 'FWD' and 'AFT' are provided to show which section of the fuselage contains fire. A stainless steel firewall divides the engine compartment at a point immediately aft of the compressor, therefore 'FWD' indicates a fire in the compressor and accessory section; 'AFT' the combustion chamber and tailpipe. The system can be tested

for integrity of the fire-detection transducers on the ground by pressing the fire test button located next to the warning lights.

Note
There is no fire extinguishing system on this airplane.

ABOVE LEFT Auxiliary nose-door parts list.

ABOVE Nose-wheel steering system schematic.

LEFT Except for the first prototype XP-86 (PU-597), no fire extinguisher was ever fitted to the F-86.

LEFT The fire-detection warning lights are prominently positioned on the right-hand side of the instrument coaming. *(Author)*

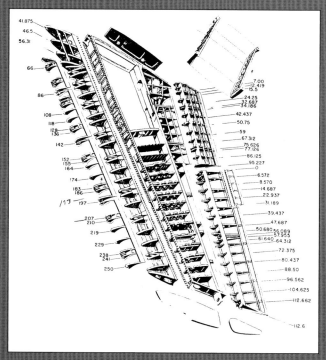

ABOVE The wing structure was completely new and (at the time) a genuine innovation. The wing had to cope with extremely high stresses yet remain thin in cross section (to minimise drag) and light enough to give the aircraft good performance; from an engineering point of view these requirements were all diametrically opposed. The solution required new construction techniques to be invented.

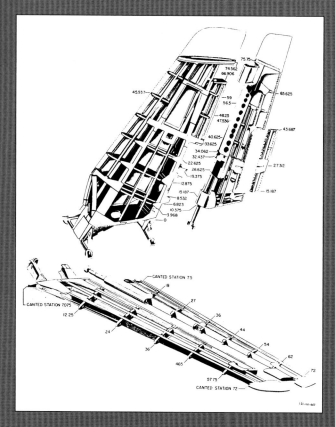

ABOVE Fin, rudder, tail-plane and elevator construction.

BELOW The fuselage construction was a relatively conventional rib and stringer structure.

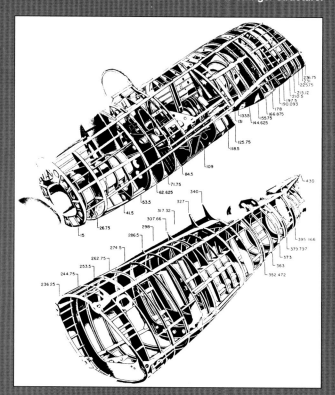

Airframe construction techniques

The Sabre's structure had to withstand not only the rigours of high-speed flight, but also extreme temperatures, especially in local areas surrounding the engine turbine and jet-pipe. The potential for high 'g' force loading was also a major consideration given that the aircraft could travel at speeds where the structural limits could easily be exceeded.

The thinness of the aircraft's wing effectively ruled out conventional rib and stringer design, and so a machine-milled double-skinned structure with 'hat sections' between the layers was devised. This method provided room for the installation of self-sealing fuel tanks within the wing structure. The wing skin thickness was also tapered throughout its length and width, being .250in thick at the wing root, tapering down to .064in at the joint with the outer wing skin, and .032in at the wing-tip. Not only did this construction method allow for a lighter structure, but also the wing skins could be more precisely tuned to cope with varying stresses of flight.

So complex were these tapered skins that they required the development of new manufacturing tools

ABOVE, LEFT AND BELOW To facilitate engine removal and access to accessories, the whole rear fuselage could be removed from a point just aft of the wing. This not only presented new structural challenges, but also created the problem of how to get all electrical wires, fire-detection loops, hydraulic pipes and flying-control connections to reliably bridge the divide. The fact that an engine could be swapped 'in the field' in a matter of hours using only basic tools is a tribute to the practical solution that was developed. *(First picture Col Pope)*

and techniques. Although the US aerospace industry was highly advanced it still did not have the necessary machining equipment for this type of construction and, therefore, NAA designed and developed the technology in-house.

If anything the Sabre was over-engineered to take account of the potential for the aircraft to break up under high wing loading and other combat stresses. This conservative approach resulted (by modern standards) in the aircraft probably being somewhat heavier than it needed to be. There is always a compromise between strength and weight, yet structural failure of the Sabre was extremely rare in spite of frequent over-stresses in action. The Sabre was indeed strong and built to last.

Surface finish of the wings and fuselage also received close attention in order to minimise drag and keep the lines of the aircraft as 'clean' as possible. In addition to some aerials being hidden behind aerodynamic fairings, all flying control hinges, fasteners and inspection panels were aerodynamically optimised. A look at the air-brake design shows just how successful these efforts were: in the closed position the air-brake panel is completely level with the lines of the tail-cone. Furthermore, flush-fitting rivets were installed across all external surfaces.

(Author)

LEFT The removable tail section features quick-connect hydraulic couplings, vent pipes and flying-control wires. *(Author)*

WARNING
THIS AIRPLANE CONTAINS A CANOPY REMOVER CONTAINING AN EXPLOSIVE CHARGE. REF. T.O. 11-1-99
WARNING
THIS AIRPLANE CONTAINS A SEAT EJECTION CATAPULT CONTAINING AN EXPLOSIVE CHARGE. SEE T.O. 11-1-98 FOR COMPLETE IN TRUCTIONS

WARNING

A ground safety pin prevents the canopy ejection gun from being fired accidentally while airplane is on the ground. (See figure 1-18.) This safety pin must be removed before flight and replaced after flight.

RIGHT An external canopy emergency release handle can be reached through an access door on the left side of the fuselage below the canopy frame. This does not fire the canopy ejection gun but simply declutches the canopy so that it can be moved manually; to assist moving the canopy a pop-out handle (visible above the yellow panel) is located on the canopy bow.

Cockpit and canopy

The electrically operated canopy may be controlled from either inside or outside the aeroplane. The canopy actuator is directly powered by the battery. Emergency release of the canopy in flight is by means of an ejection gun which fires the canopy off the aircraft. This is achieved by raising the right handgrip on the seat in preparation for seat ejection. When the canopy is ejected in flight it pulls a safety pin from the seat catapult-firing mechanism, which prevents seat ejection until after the canopy leaves the aeroplane.

A canopy declutch handle at the bottom of the centre pedestal is for ground use only and when pulled disengages the canopy from the drive shaft so it can be moved manually. The aircraft can be flown with the canopy open at up to 215 knots.

(Author)

LEFT AND BELOW Most 'A' model aircraft (and some early 'Es') were fitted with a complex 'vee' windscreen, which incorporated thick bullet-proof glass. Very early 'A's had curved windscreens.

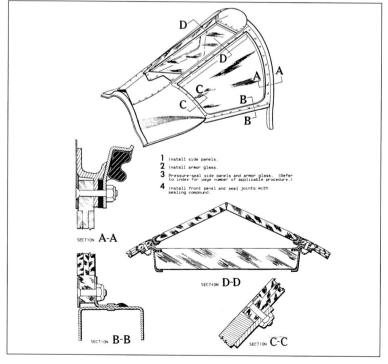

1 Install side panels.
2 Install armor glass.
3 Pressure-seal side panels and armor glass. (Refer to index for page number of applicable procedure.)
4 Install front panel and seal joints with sealing compound.

SECTION A-A

SECTION D-D

SECTION B-B

SECTION C-C

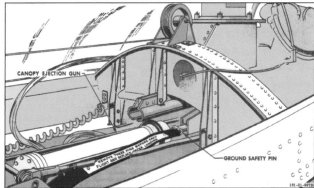

Ejection seat

An ejection seat is provided that will catapult the pilot clear of the tail surfaces, thus making bail-out safe at any speed. A catapult mounted aft of the seat supplies the propelling force to eject the seat and pilot from the cockpit. Armrests and footrests on the seat are fixed, but the handgrips provided are hinged to pull up into a vertical position for ejection. Safety pins are provided to prevent inadvertent firing of the seat and canopy on the ground.

When the seat is ejected the anti-'g' suit, oxygen hose and microphone/headset connections automatically disconnect at a single fitting attached to the seat between the footrests.

The seat catapult trigger lever is located

beneath the folding handgrip at the forward end of the right armrest and is protected by a guard and safety wire. Raising the handgrip fully breaks the safety wire and fires the canopy remover, then squeezing the exposed trigger fires the seat catapult.

The downside to maintaining any ejection seat, especially one from such an early era, is that spare

ABOVE LEFT AND RIGHT The canopy ejection gun 'disposes' of the canopy by firing it rearward along the canopy guide tracks. *(Author)*

LEFT The Sabre's electrically sliding canopy in the fully open position. In an ejection the canopy is first jettisoned by an explosive charge before the seat can be fired. *(Author)*

RIGHT AND BELOW This seat fitted to 48-178 is from a Japanese F-86F licence-built by Mitsubishi. *(Author)*

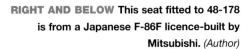

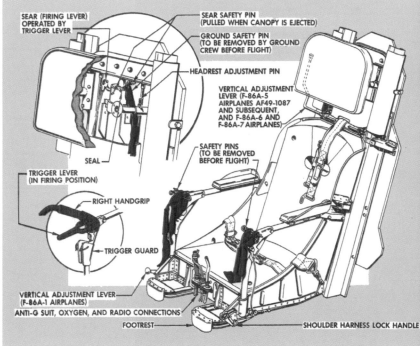

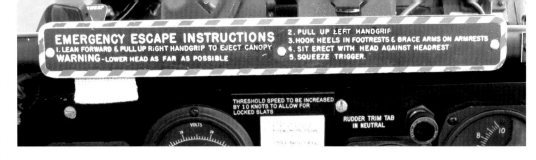

RIGHT Ejection seat 'instructions for use' are mounted directly in front of the pilot. It is hoped that there will be time to read them.

(All photographs this spread author)

BELOW Air conditioning, defrosting, windshield anti-icing and pressurisation controls are located aft of the throttle quadrant. The emergency air-conditioning system/ smoke-removal plan is to open the canopy!

parts are very difficult to source and then incredibly expensive, this being especially true of the seat and canopy pyrotechnics. US government regulations concerning the export of explosives makes for a seat overhaul that is unbelievably expensive (not helped by there being a supplier monopoly) and mired in a trail (and trial) of seemingly endless paperwork. The maintenance issues related to the ejection seat create one of the major engineering challenges with this aircraft, yet it is not something that can ever be compromised on.

Operational equipment

Other items of operational equipment include: oxygen, communications, cockpit pressurisation, heating and ventilating, pitot heat, anti-'g' suit and lighting.

Air conditioning and pressurisation

The cockpit of the F-86 was fully pressurised and air conditioned utilising air from the engine's 11th compressor stage at a rate of 10lb per minute at 'full cool' or 14lb per minute in the 'full heat' setting. Air was routed either through a turbine refrigeration unit for cold settings or bypassed for hot.

The canopy/windscreen defrosting system had to be capable of dealing with the rapid

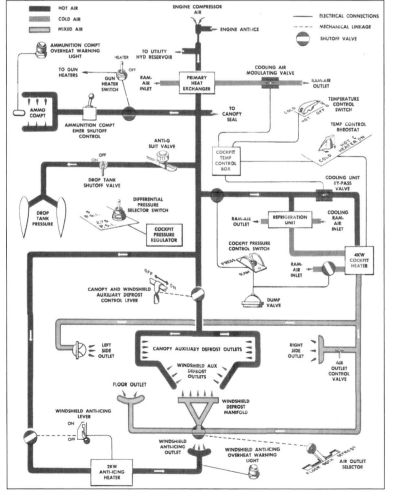

descent rates from altitude of which the Sabre was capable. The system performs very well, unlike contemporary Russian designs: it is known that several MiG-15s were shot down in Korea simply because their pilots could not see out through their frosted canopies.

Instruments

Most of the instruments are (AC) electrically powered directly from the battery via the static inverter. The tachometer and exhaust temperature gauge are self-generated from their

LEFT The airspeed indicator is conventional, with the addition of a maximum airspeed-indicating mechanism which automatically indicates maximum allowable airspeed for the existing flight altitude. A fluorescent pointer shows indicated airspeed and a red pointer shows the maximum allowable airspeed. The indicator is preset for the limiting Mach number of the aeroplane and the red pointer moves to indicate the airspeed corresponding to the limiting Mach number at the existing flight altitude. There is also a separate Mach meter.

LEFT Other instruments include a conventional accelerometer (that shows positive accelerations only), a gyro compass (the compass card is rotated by the pilot and needle shows the aircraft's heading), altimeter and a type A-1 attitude gyro.

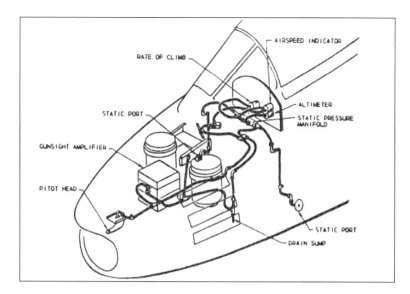

ABOVE This diagram of an early F-86A shows the location of the pitot tube inside the intake; due to errors caused by pressure fluctuations inside the intake this was repositioned to the right wing-tip on later aircraft.

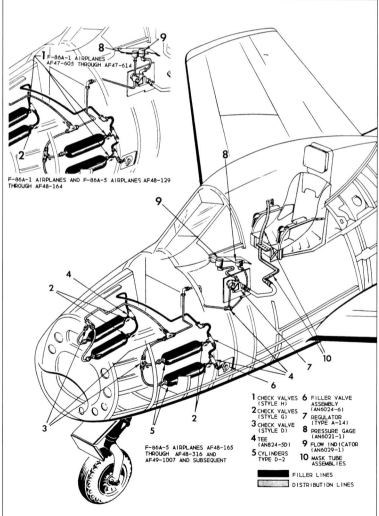

F-86A-1 AIRPLANES AF47-605 THROUGH AF47-614

F-86A-1 AIRPLANES AND F-86A-5 AIRPLANES AF48-129 THROUGH AF48-164

F-86A-5 AIRPLANES AF48-165 THROUGH AF48-316 AND AF49-1007 AND SUBSEQUENT

1 CHECK VALVES (STYLE H)
2 CHECK VALVES (STYLE G)
3 CHECK VALVE (STYLE D)
4 TEE (AN824-5D)
5 CYLINDERS TYPE D-2
6 FILLER VALVE ASSEMBLY (AN6024-6)
7 REGULATOR (TYPE A-14)
8 PRESSURE GAGE (AN6021-1)
9 FLOW INDICATOR (AN6029-1)
10 MASK TUBE ASSEMBLIES

■ FILLER LINES
▨ DISTRIBUTION LINES

own systems. Fuel, hydraulic and oil pressure gauges are direct pressure reading instruments. An electric (DC) automatic vibrator is mounted on the forward side of the instrument panel to reduce instrument lag or sticky pointer indications.

Oxygen system

The low-pressure oxygen system is supplied from four Type D-2 cylinders installed in the nose. For combat safety, check valves are incorporated in the system. The complete system utilises a single-point refilling valve located on the left side of the fuselage forward of the wing. Included in the system is a pressure-demand regulator, a flow indicator and a pressure gauge. Normal minimum pressure is 400psi.

ABOVE AND LEFT The oxygen system engineering drawing shows the location of the oxygen bottle. For weight and balance reasons on 48-178 these have been relocated to the area once occupied by the guns. (Author)

CABIN ALT - FEET -	GAGE PRESSURE - PSI							
	400	350	300	250	200	150	100	BELOW 100
40,000	5.7	4.9	4.1	3.2	2.4	1.6	0.8	
	5.7	4.9	4.1	3.2	2.4	1.6	0.8	
35,000	5.7	4.9	4.1	3.2	2.4	1.6	0.8	
	5.7	4.9	4.1	3.2	2.4	1.6	0.8	
30,000	4.2	3.6	3.0	2.4	1.8	1.2	0.6	
	4.2	3.6	3.0	2.4	1.8	1.2	0.6	
25,000	3.4	2.9	2.4	1.9	1.4	1.0	0.5	
	4.0	3.4	2.8	2.3	1.7	1.1	0.6	
20,000	2.7	2.3	1.9	1.5	1.2	0.8	0.4	
	4.5	3.9	3.2	2.6	1.9	1.3	0.6	
15,000	2.1	1.8	1.5	1.2	0.9	0.6	0.3	
	5.4	4.6	3.9	3.1	2.3	1.5	0.8	
10,000	1.8	1.5	1.3	1.0	0.7	0.5	0.3	
	7.2	6.2	5.2	4.1	3.1	2.1	1.0	

Black figures indicate diluter lever "NORMAL."
Red figures indicate diluter lever "100%."
Cylinders: Four Type D-2

LEFT The table shows that oxygen duration from a fully charged system (assuming no leaks) would greatly exceed the endurance of the aircraft.

Chapter Six

The Engineer's View

An engineer once described the F-86 as a 'spanners and hammers aircraft: big spanners and big hammers: oh, and a screwdriver would be handy too!' This elegantly describes an aircraft that is robustly engineered and can be kept going with the most basic of tools. It's good old-fashioned engineering: there's nothing dainty or fragile about the Sabre, which given its intended purpose is unsurprising.

OPPOSITE 48-178 undergoes maintenance in the winter of 2010. All maintenance services for the aircraft are provided by the Duxford-based Aircraft Restoration Company. *(Author)*

Maintaining the World's oldest flying Sabre

The aircraft is actually quite simple and straightforward to maintain or repair as was proved time and again when it was in service. The engineer needs to apply his or her skills to accomplish basic engineering tasks: the art is to know when and how to use brute force. Some procedures are a challenge, notably those associated with the hydraulics where the space available to work is always just a fraction too small.

A great engineering advantage is that the aircraft can be split in two, thereby revealing the engine and jet-pipe. Most of the other systems (starter/generator, plus numerous pumps, fuel controllers, filters and valves) are bolted to the front of the engine and accessible by crawling down the intake.

What follows is a photographic record of some of the mostly preventative or scheduled maintenance procedures carried out over the last few years by the Aircraft Restoration Company at Duxford on F-86A 48-178. In service the Sabre was very reliable. In private hands the story is similar, but at over sixty years old the machine needs to be carefully managed to keep it in top-notch condition.

Spares

Original OEM spare parts for the Sabre are surprisingly abundant and are mostly sourced from the USA, where vast quantities still exist. Golden Apple Operations have large stocks of spare parts too, especially pumps, valves, filters and consumables such as tyres and brake pads. Almost everything that comes off the aircraft is sent away for overhaul and then put back on the shelf. The opportunity to obtain spares is never overlooked.

A study of the parts lists also reveals that many components were common across all Sabre variants and, given the considerable number of Sabres produced, this adds to the possibilities. Many Sabres have survived in museums and occasionally bits and pieces can be liberated from such sources, sent for overhaul and then put on the shelf.

Routine servicing

For land-aways the pilots are trained to do a limited amount of servicing, principally to check and replenish fluids. Thankfully, the aircraft is simple to service in this respect. When flying from the aircraft's home base at Duxford all servicing and pre-flight preparation is carried out by the Aircraft Restoration Company.

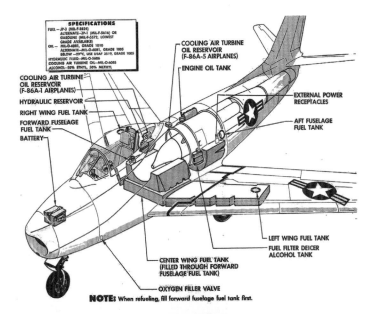

SPECIFICATIONS
FUEL — JP-3 (MIL-F-5624)
ALTERNATE—JP-1 (MIL-F-5616) OR
GASOLINE (MIL-F-5572, LOWEST
GRADE AVAILABLE)
OIL — MIL-O-6081, GRADE 1010
ALTERNATE—MIL-O-6081, GRADE 1005
BELOW —39°C, USE USAF 2519, GRADE 1005
HYDRAULIC FLUID-MIL-O-5606
COOLING AIR TURBINE OIL—MIL-O-6085
ALCOHOL—50% ETHYL, 50% METHYL

COOLING AIR TURBINE
OIL RESERVOIR
(F-86A-5 AIRPLANES)

ENGINE OIL TANK

COOLING AIR TURBINE
OIL RESERVOIR
(F-86A-1 AIRPLANES)

HYDRAULIC RESERVOIR

RIGHT WING FUEL TANK

FORWARD FUSELAGE
FUEL TANK

BATTERY

EXTERNAL POWER
RECEPTACLES

AFT FUSELAGE
FUEL TANK

LEFT WING FUEL TANK

FUEL FILTER DEICER
ALCOHOL TANK

CENTER WING FUEL TANK
(FILLED THROUGH FORWARD
FUSELAGE FUEL TANK)

OXYGEN FILLER VALVE

NOTE: When refueling, fill forward fuselage fuel tank first.

FAR LEFT AND LEFT The Sabre has six refuelling points (forward fuselage, rear fuselage, two wings and two drop tanks). All are open-line gravity fillers.
(Col Pope)

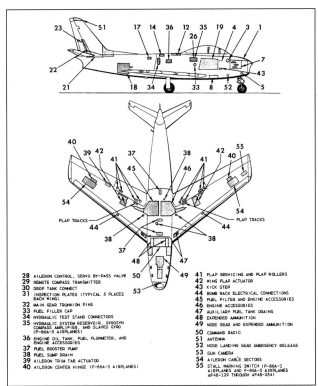

LEFT AND BELOW F-86A access and inspection provisions are numerous.

1 RADAR, BATTERY, AND GUN SIGHT
2 OXYGEN FILLER VALVE
3 HYDRAULIC EQUIPMENT (F-86A-5 AIRPLANES)
4 CANOPY CONTROL SWITCHES
5 NOSE GEAR TRUNNION PIN AND OXYGEN BOTTLES SUPPLY LINE FITTINGS (F-86A-1 AIRPLANES)
6 GUNS, GUN CHARGER, BORE SIGHT FIXTURE, CABIN PRESSURE CHECK, AND AMMUNITION LOADING HOOK
7 ELECTRICAL, OXYGEN, AND ENGINE LINE DISCONNECTS
8 AMMUNITION BOXES
9 GROUND FIRE FIGHTING, DATA CASE, AND ENGINE LINE DISCONNECTS
10 EMERGENCY CANOPY RELEASE
11 RADIO, FUEL LEVEL CONTROL SWITCH
12 HEAT AND VENT ADJUSTMENT AND COOLING UNIT OIL FILLER (F-86A-5)
13 SURFACE CONTROL ADJUSTMENT
14 FUEL AND OIL PRESSURE TRANSMITTERS (F-86A-5 AIRPLANES AF48-225 AND SUBSEQUENT)

15 EXTERNAL POWER CONNECTION
16 FUEL VENT
17 SPARK PLUGS AND ENGINE DISCONNECTS
18 SPEED BRAKES
19 GUNS, GUN CHARGER, AMMUNITION LOADING HOOK
20 CONTROL CABLES AND PULLEYS
21 ELEVATOR HYDRAULIC BOOST UNIT
22 HORIZONTAL STABILIZER ACTUATOR
23 RUDDER TRIM TAB ACTUATOR
24 ANTENNA DISCONNECT PLUG
25 BRAKE RESERVOIR (F-86A-1 AIRPLANES AND F-86A-5 AIRPLANES AF48-129 THROUGH AF48-316)
26 COOLING UNIT OIL FILLER (F-86A-1 AIRPLANES)
27 DE-ICING TANK FILLER (F-86A-5 AIRPLANES AF48-256 AND SUBSEQUENT)

28 AILERON CONTROL, SERVO BY-PASS VALVE
29 REMOTE COMPASS TRANSMITTER
30 DROP TANK CONNECT
31 INSPECTION PLATES (TYPICAL 5 PLACES EACH WING)
32 MAIN GEAR TRUNNION PINS
33 FUEL FILLER CAP
34 HYDRAULIC TEST STAND CONNECTORS
35 HYDRAULIC SYSTEM RESERVOIR, GYROSYN COMPASS AMPLIFIER, AND SLAVED GYRO (F-86A-5 AIRPLANES)
36 ENGINE OIL TANK, FUEL FLOWMETER, AND ENGINE ACCESSORIES
37 FUEL BOOSTER PUMP
38 FUEL SUMP DRAIN
39 AILERON TRIM TAB ACTUATOR
40 AILERON CENTER HINGE (F-86A-5 AIRPLANES)

41 FLAP SERVICING AND FLAP ROLLERS
42 WING FLAP ACTUATOR
43 KICK STEP
44 BOMB RACK ELECTRICAL CONNECTIONS
45 FUEL FILTER AND ENGINE ACCESSORIES
46 ENGINE ACCESSORIES
47 AUXILIARY FUEL TANK DRAINS
48 EXPENDED AMMUNITION
49 NOSE GEAR AND EXPENDED AMMUNITION
50 COMMAND RADIO
51 ANTENNA
52 NOSE LANDING GEAR EMERGENCY RELEASE
53 GUN CAMERA
54 AILERON CABLE SECTORS
55 STALL WARNING SWITCH (F-86A-1 AIRPLANES AND F-86A-5 AIRPLANES AF48-129 THROUGH AF48-254)

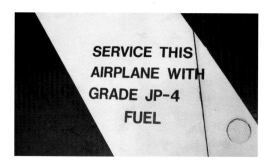

ABOVE The modern (civilian) equivalent to JP-4 is Jet A-1. *(Author)*

ABOVE The oil tank is refilled by gravity and as there is no oil gauge a 'dip-stick' is incorporated into the tank to check the level. The tank should not be overfilled; on shutdown scavenging can cause oil to seep into the hot parts of the engine and then causes white smoke to pour from the jet-pipe, which could potentially catch alight with obvious consequences. *(Author)*

ABOVE Pilots are trained and approved to carry out basic servicing such as refuelling and the replenishment of oil and hydraulic fluid. One way to check if the drop tanks are empty is to tap them and then listen for a hollow echo. *(Author)*

RIGHT Whenever the aircraft is parked outside it is always protected from the elements with a purpose-made canopy/intake cover. *(Author)*

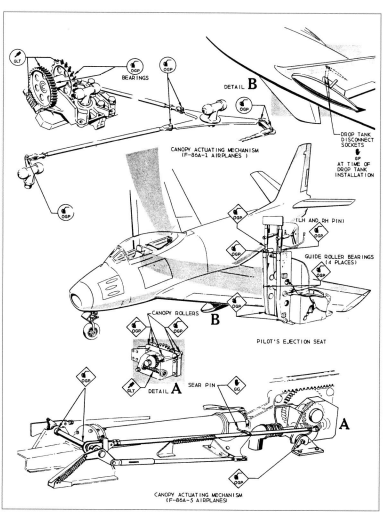

ABOVE Ejection seat and canopy 'moving parts' require regular lubrication.

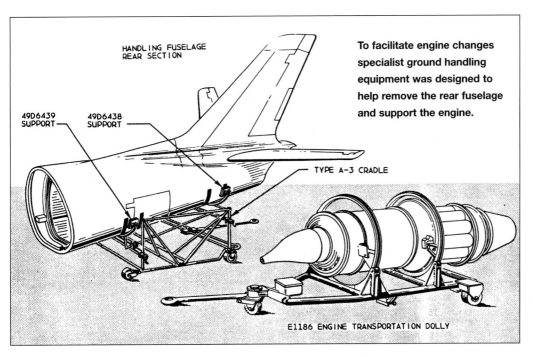

HANDLING FUSELAGE
REAR SECTION

49D6439
SUPPORT

49D6438
SUPPORT

TYPE A-3 CRADLE

E1186 ENGINE TRANSPORTATION DOLLY

To facilitate engine changes specialist ground handling equipment was designed to help remove the rear fuselage and support the engine.

Engine/tail refitting

The following sequence (this page and overleaf) shows the refitting of the jet-pipe, followed by the rear fuselage. Both jobs require teamwork, a steady pair of hands, a shoulder to lean on and a careful systematic approach – this is not the time to drop anything.
(Col Pope)

Refitting the canopy and ejection seat

The canopy has to be removed to allow the ejection seat to be taken out and refitted for servicing. Removal and refitting is a team job – this is one component that would almost certainly not survive being dropped from a great height. *(Author 1–5; Col Pope 6–10)*

Installing and adjusting the cockpit canopy.

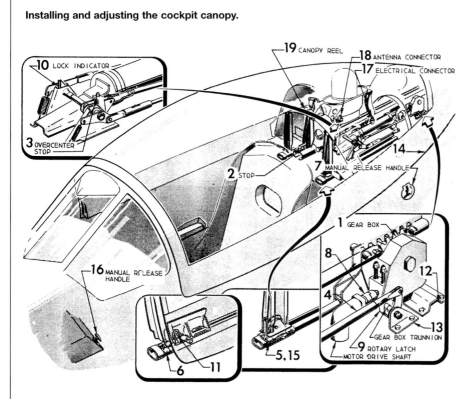

19 CANOPY REEL
18 ANTENNA CONNECTOR
17 ELECTRICAL CONNECTOR
10 LOCK INDICATOR
3 OVERCENTER STOP
2 STOP
7 MANUAL RELEASE HANDLE
14
16 MANUAL RELEASE HANDLE
1 GEAR BOX
8
12
4
13
5,15
6
11
9 ROTARY LATCH
MOTOR DRIVE SHAFT
GEAR BOX TRUNNION

1,2 Install gear box on canopy at forward end of rear track, and install stop after gear box has been manually positioned aft by rotating drive shaft to enable installation of stop.

3 Adjust overcenter stop (by varying washer arrangement) to align lock indicator pointers.

4 Adjust rotary latch push-pull rods so that slots of rotary latches are aligned vertically when linkage is overcenter, against stop.

5,6 Install canopy, engaging trucks at aft end of both intermediate and forward tracks.

7 Open rotary latches by pulling either inside or outside manual release handle.

8,9 Engage motor drive shaft with gear box drive shaft and rotary latches with gear box trunnions.

10 Release manual handle, and verify proper position of lock indicator.

NOTE: Steps 11, 12, 13, and 15 are to be followed during installation of new canopy, but are unnecessary when reinstalling original canopy.

11 Using washers as necessary, adjust canopy at forward trucks to align with windshield.

12 Using shims as necessary, accomplish proper support of gear box so that gear box drive shaft is aligned with motor drive shaft and aft "free play" of canopy is minimized.

13 Using shims as necessary, adjust rotary latch brackets to align latches properly with gear box trunnions.

14 Check canopy skirt for 1/16-inch minimum clearance in closed position.

15 With eccentric washer, adjust trucks at intermediate track to move freely within track.

16 Close canopy, pull inside manual release, and, from inside airplane, manually check for free movement of canopy along tracks, observing that canopy is clear of inflatable seal. If canopy movement is difficult, repeat steps 12 through 16 as necessary.

NOTE: Pressure-seal bolt heads in deck where gear box support and rotary latch brackets are attached; then check cockpit pressurization. (Refer to index for page number of applicable procedure.) If proper pressurization cannot be maintained, repeat steps 11, 12, 13 and 15 through 17.

17,18 Connect canopy electrical wiring connector and plug in antenna connector.

19 Engage canopy reel cable in quick-disconnect fitting.

Flying controls

All flying-controls surfaces (elevator, rudder, flaps, speed-brakes and ailerons) are removed for non-destructive testing and lubrication. Refitting is another effort requiring precision teamwork. Elevators and rudder have been removed for inspection, the replacement of worn bearings and lubrication of moving parts. The refitting requires a head for heights and a couple of pairs of steady hands.

(Col Pope)

(Mark Linney)

LEFT AND BELOW Quick connectors for flying controls, hydraulics and electrical connections allow the rear fuselage to be quickly removed and re-joined.

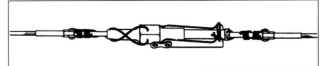

BELOW LEFT AND RIGHT The wings are attached to the centre-wing box section by multiple bolts which are regularly checked for corrosion. *(Author)*

BOTTOM LEFT AND RIGHT With the rear fuselage removed the combustion chamber cans of the J47 are easily accessed. *(Author)*

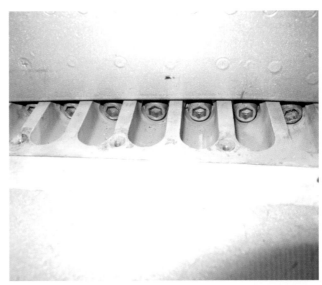

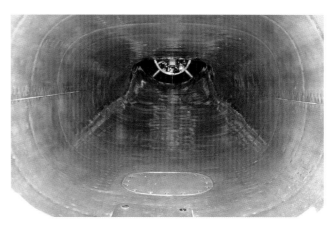

RIGHT AND BELOW The 'hot-end' of the engine can be closely inspected with the rear fuselage, jet-pipe and engine tail-cone removed. The turbine disc is subject to high levels of thermal and rotational stresses. The condition of this component is largely what determines the 'life' of the engine. *(Author)*

ABOVE LEFT AND ABOVE The accessory gearbox for the hydraulic pump and electrical starter motor/generator is mounted on the front of the engine. In use an aerodynamic fairing nose-cone covers these items to ensure a smooth flow of air to the compressor and also provide gearbox cooling. Servicing of these components along with the fitting of the fairing involves crawling down the intake. In the second to last picture the inspection cover at the bottom of the intake has been removed to allow access to the nose landing-gear pivot point for examination and lubrication. *(Author)*

BELOW The turbine disc and blades require frequent and careful inspection and the opportunity for this is taken whenever the fuselage is split. *(Col Pope)*

LEFT Andrew Foster of ARC works on the electrical wiring harness that connects to the circuit-breaker panel the other side of this bulkhead. *(Simon Condie ARPS LBIPP)*

ABOVE Access to the power control cables is significantly easier with the ejection seat removed.

ABOVE, LEFT AND RIGHT New wheel-bearing kits are fitted to the main wheels during winter maintenance along with fresh tyres. *(Author)*

Technical data and servicing requirements (for F-86A-7)

Engine	General Electric J47-GE-7 – thrust 5,200lb
Span	37ft 1in
Length	37ft 6in
Height	14ft 9in
Wheel tread	8ft 4in
Empty weight	10,854lb
Max. take-off weight	17,250lb

Operational Requirements
(Notes to Visiting Aircraft Handlers)

Fuel type	Jet A-1 AVTUR. Gravity open-line refuelling
Capacities	Internal 435 US gallons; external two 120 US gallon drop tanks. Total usable fuel 675 US gallons
Filling sequence	Forward fuselage, then rear fuselage, then wing tanks, then drop tanks as indicated on filler caps
Engine oil	Aero Shell Turbine Oil 2. Oil tank and dip-stick located behind servicing panel on right-hand side of fuselage midpoint of wing chord
Hydraulic fluid	OM15 or US spec Mil-H-5606 – 2 x hydraulic accumulators to be charged with nitrogen to 1,200psi; charging point in nose-wheel bay. Hydraulic tank and sight glass located behind servicing panel on right-hand side of fuselage forward of oil servicing panel
Electrical	28V DC required outputting 1,600A surge, 600A continuous. Three-pin standard connection behind panel on left-hand side of fuselage above wing trailing edge
Fire cover	Advisable for engine start and shut-down
Oxygen	Gaseous oxygen with pressure gauge located inside cockpit. Charging point is located behind a dedicated servicing panel on left-hand side of nose
Towing	A dedicated tow-bar is located in a compartment aft of the ammunition bays under the fuselage and is accessible from underneath the left-hand wing root leading edge. The nose-wheel steering system must be disconnected before towing. The aircraft does not have a functional parking brake and must be wheel-chocked if left unattended
Flying control lock	Internal
Ejection seat and canopy remover (safe for parking)	When parked, one safety pin in each arm rest and one pin in each of the two initiators (behind seat-back). Normal canopy opening is by electric motor. External operating open/close push buttons are located on either side of the fuselage below the windscreen quarter lights. Internal switch is located on the upper left cockpit coaming

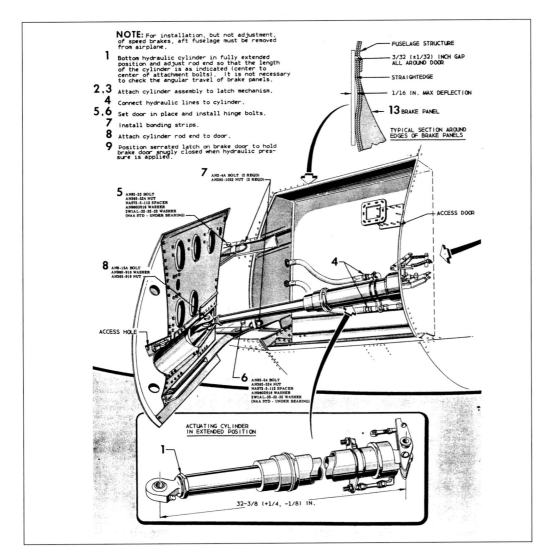

NOTE: For installation, but not adjustment, of speed brakes, aft fuselage must be removed from airplane.

1. Bottom hydraulic cylinder in fully extended position and adjust rod end so that the length of the cylinder is as indicated (center to center of attachment bolts). It is not necessary to check the angular travel of brake panels.

2, 3. Attach cylinder assembly to latch mechanism.

4. Connect hydraulic lines to cylinder.

5, 6. Set door in place and install hinge bolts.

7. Install bonding strips.

8. Attach cylinder rod end to door.

9. Position serrated latch on brake door to hold brake door snugly closed when hydraulic pressure is applied.

— FUSELAGE STRUCTURE
— 3/32 (±1/32) INCH GAP ALL AROUND DOOR
— STRAIGHTEDGE
— 1/16 IN. MAX DEFLECTION
13 BRAKE PANEL

TYPICAL SECTION AROUND EDGES OF BRAKE PANELS

7 AN3-4A BOLT (2 REQD)
AN365-1032 NUT (2 REQD)

5 AN85-23 BOLT
AN365-524 NUT
NAS72-5-112 SPACER
AN960D516 WASHER
2W1AL-32-32-32 WASHER
(NAA STD - UNDER BEARING)

8 AN9-15A BOLT
AN960-916 WASHER
AN365-916 NUT

ACCESS HOLE

4

ACCESS DOOR

6 AN85-24 BOLT
AN365-524 NUT
NAS72-5-112 SPACER
AN960D516 WASHER
2W1AL-32-32-32 WASHER
(NAA STD - UNDER BEARING)

ACTUATING CYLINDER IN EXTENDED POSITION

1

32-3/8 (+1/4, -1/8) IN.

LEFT Instructions for installing and adjusting speed-brake panels.

BELOW AND RIGHT When a thin sheet stainless-steel shroud surrounding the jet-pipe was cracked and in need of specialist welding it was fortunate that the services of the Welding Institute (TWI), located close to Duxford, could be engaged. The purpose of the shroud is to deflect any leaking exhaust gases during the brief period between start-up and the engine reaching running temperature, when previously weeping seals around the jet expand and become leak proof. Although within limits, several small cracks and tears were in need of attention. Using TIG welding, and an appropriate consumable for the aged material, TWI was able to arrest and repair cracks in the wafer-thin material. *(Author/Simon Condie ARPS LBIPP)*

The Sabre in Korea – a Crew Chief's recollection

Norm Kalow entered the USAF immediately after high school, completed the aircraft and engine mechanic's training course at Sheppard Field, Texas, and then joined the 4th Fighter Group at Langley Field, Virginia. Norm was a member of the original contingent that deployed to Korea from the USA in November 1950.

Norm gives his appraisal of the F-86 during Korean operations:

Generally, the Sabre was a great aircraft. Most repairs were easy as compared to reciprocating engine aircraft (except for a few items, such as canopy seals), and we experienced minimal servicing problems at the primitive air fields and facilities in Korea. As an example, two of us could change an engine, including the engine's run-up, at a normal working pace in under an hour with few tools. All a good crew chief needed was a screwdriver and a pair of water pump pliers, and we could fix 90% of the problems!

Inactivity was bad for the Sabre: the less it flew, the more minor problems it seemed to have.

Initially at Johnson Air Base, shortly after we brought the Sabres up to flight status after the rigors of the ocean crossing from California, many small problems surfaced. Foremost were the fuel level transmitters in the wing cells. To change them required draining the fuel and dropping the leading edge slats. This was a time-consuming procedure, but not particularly difficult. After the Sabres got to Korea and their flying increased, many other problems eliminated themselves.

One major problem I recall was the rupturing of a rubber membrane in the engine fuel flow control valve. True to Murphy's Law, this was usually discovered after the last flight of the day when we were checking the engine oil reservoir. Upon removing the cap, a mixture of JP-1 fuel and oil would gush out, accompanied by an oath from a crew chief! Repairs consisted of pulling the engine, replacing the fuel flow control valve, and then putting everything back together. But we found that the attention to the maintenance and repair function was extremely well thought out in the Sabre by its designers and engineers, such as in the use of electrical 'cannon' plugs, quick disconnect hydraulic lines, and simple mechanical methods for securing components. As a further example, the mounting of the J47 jet engine was secured by two ball and sockets and a forward cross pin. Elegant!

The worst repair I recall in Korea was replacing the Sabre's canopy seal as it was glued into a channel. It had to be ripped out and scraped, a new seal glued in its place, and then flight tested by a pilot at altitude. The seal was hollow inside and inflated by engine compressor bleed air to ensure a tight seal. It required a 'few trial and error' gluing sessions to get it right.

For such a comparatively complex aircraft (for its time), the Sabre was amazingly trouble

RIGHT Most 'field' maintenance carried out at Kimpo Air Base was done in the open air. This made the job especially arduous in the winter months. Fortunately the Sabre was reasonably reliable and for the most part straightforward to work on. *(USAF)*

free. This also included its systems, such as the hydraulics, the electricals and so forth.

The Sabre had a small 'footprint', and due to its weight, if it went off the concrete or pierced steel planking, it got stuck and required a small tractor to be pulled back to solid footing. But thanks to the hydraulic power steering in the nose gear, the pilot usually had good control when 'driving' the Sabre on the ground.

My impression of the Sabre 40 years later is of a solid, well-designed, no nonsense airplane, with a characteristic beauty and form all its own. On the ground or in flight, with drop tanks or 'clean', the F-86 looked 'classy'. I can still remember watching a returning flight of four Sabres in a trailing formation making a low pass over the field and peeling off, one by one, into the final landing circle. Beautiful! And in its landing attitude a semi-stall, nose-high configuration; it looked like no other jet-fighter we had at our base. The Sabre was the 'cock of the walk', and it landed like it darn well knew it!

All in all, the Sabre was docile, but if we were careless, it would 'bite'. A case in point was of a young crew chief: cocky, slightly careless, and sometimes unheeding of advice. In early 1951, an advance contingent of Sabres was at K-14 (Kimpo), Korea. A Sabre's engine was run up on the flight line when the young man walked in front of the air intake duct. He was instantly sucked in and rammed through twelve feet of intake duct to the engine where the engine screens stopped him. He was dead by the time the engine was shut down and he was pulled out.

We were always warned about the inherent dangers of the nose intake, and we treated it with great respect. But there was no problem with the Sabre's exhaust. People would be knocked down if they got too close, but not burnt or seriously hurt. The local Koreans hired to sweep our taxi-ways sometimes would not stay clear of the jet exhausts, and we periodically saw Koreans tumbling rear over tea kettle with their white clothes flying if they did not respect the Sabre's powerful exhaust.

A more serious problem was making sure that saboteurs did not throw sand or gravel into our Sabre's intake ducts. We heard they were paid a small sum by the 'other side' to do this. This would cause serious engine problems if they were successful.

Speaking of engines, the Sabre's J47 jet engine required more care in 'firing up' than say the F-80 Shooting Star, or else a 'hot start' resulted, with a large tongue of flame shooting out the tailpipe. This could ruin the engine. We crew chiefs had to be on the ball whenever we ran the engines up, but fortunately the learning curve was not excessively long. As I recall, during engine start-up, as the rpm came up, careful manipulation of the throttle and a watchful eye on the fuel pressure gauge until the tailpipe temperature gauge 'came alive' resulted in a safe, smooth start-up. The key point was a controlled low fuel pressure reading via the throttle, or else we could sizzle someone's shorts a hundred feet away with the flame blast!

The Sabre was rugged, and a gear-up landing was usually repairable. I recall one new pilot, just over from the States, who landed a new Sabre with its gear up. He was just forgetful. He said he could not hear the tower warning him on the radio because of a loud noise in the cockpit. It was his gear-up warning horn! The crew chief of that Sabre, a corporal, had the pilot, a first lieutenant, as his helper until that plane was repaired!

ABOVE F-86A was a 'crew chief's dream'. Here a Sabre of the 4th FIW undergoes an engine change in the field, an activity that could be completed by two men in under an hour.

Appendix 1

Sabre survivors

With almost 10,000 Sabres and Furys produced it is unsurprising that, in one form or another, hundreds remain intact and preserved; unfortunately, very few remain airworthy. Given US policy to put ex-military aircraft 'beyond use', most of those that do still fly are aircraft that were either exported or are non-US built from Canada or Australia. An exception to this is 48-178.

Many retired airframes continued to serve as decoys or firefighter, maintenance and battle-damage repair-training airframes. Some lie derelict with uncertain fates. Dozens are in museums and many others sit upon poles at the entrance to air force bases worldwide.

BELOW F-86F (51-13407) preserved as gate-guardian at PAF Base Lahore, Pakistan. *(A. Hussain)*

One or two have even been 'sectioned' and are displayed for educational purposes. Even countries that never operated the F-86 seem duty bound to include an example of the type somewhere in their museum collections; such is the importance of the aircraft in the story of flight.

For some Sabres there are political undertones to their preservation. For example, on 1 June 1963 Tingze Xu, a pilot in the Republic of China Air Force 43 Squadron, defected to the People's Republic of China in an F-86F (serial number F-86272/52-4441). This aircraft now resides in Beijing's Museum of the People's Revolution. Similarly, wreckage (wing sections, twisted 0.50-calibre machine guns, rocket ordnance, canopies, a burnt-out jet-engine and ejection seats) of Pakistan Air Force Sabre F-86F No. 52-5248 can be viewed at the Indian Air Force Museum in Palam, New Delhi. This is one of two PAF Sabres at the museum shot down over Halwara by IAF Hawker Hunters. Likewise, the People's Republic of Korea (North Korea) displays the shot-down wreckage of Republic of Korea (South Korea) F-86D No. 52-10031 at the PRK Army Museum at Pyongyang.

Many Sabres are displayed in the colours of national display teams, which almost every country to operate the aircraft fielded at some point during service. In the case of Australia, Canada, Italy and the USA more than one team chose the F-86 as their mount.

Appendix 2

Principal production Sabre variants

F-86A

The original, purest, lightest and least complex Sabre; 554 were built.

F-86D

The all-weather/night-fighter interceptor development of the Sabre. Fitted with a J47-GE-17 with afterburner rated at 5,425lb thrust. Gun armament was eliminated in favour of a retractable under-fuselage tray carrying 24 unguided Mk 4 rockets, then considered a more effective weapon against enemy bombers than a barrage of cannon fire.

The fuselage was wider and the airframe length increased to 40ft 4in, with clam-shell canopy, enlarged tail surfaces and AN/APG-36 all-weather radar fitted in a radome in the nose, above the intake. Later models of the F-86D received an uprated J47-GE-33 engine rated at 5,550lb. A total of 2,504 D-models were built.

F-86E

The improved day-fighter model that followed the F-86A. It was ordered on 17 January 1950, and serial number 50-579 first flew on 23 September 1950 with George Welch at the controls. A total of 456 'E' Sabres were built with deliveries to the Air Force beginning on 9 February 1951. The new Sabre was first allocated to the Air Defense Command's 33rd Fighter Interceptor Wing by May 1951. The 'E' was sent to Korea by July 1951.

The F-86E was powered by a General Electric J47-GE-13 turbo-jet rated at 5,200lb of thrust, the same as the last F-86A, but due to the extra weight the service ceiling fell to 47,200ft. Mechanical engine control remained the same with the major difference over the 'A' being full hydraulic irreversible controls and a much improved horizontal tail-control design with artificial feel: there were no longer any trim tabs (except the rudder, which used the same conventional cable system with an electrically adjusted trim tab as the 'A'). These elements made what NAA referred to as 'super controls'. An artificial feel system was added as air loads were no longer transmitted to the control column. The artificial feel system worked by an arrangement of spring bungees attached to the controls. These applied loads according to the degree of control-stick deflection. The trim switch on the top of the stick actually repositioned the bungees to a different load-free spot.

Instead of only the elevator controlling the Sabre, the entire horizontal tail section now moved. The change provided improved control for the F-86E through the transonic speed range and it eliminated control reversal caused by shock-wave formation on the elevator at high Mach number. The all-flying tail combined the horizontal stabiliser and the elevators into one unit.

Externally the two models were similar, but the 'E' added a raised aerodynamic fairing at the base of the vertical fin and rudder above the all-flying tail. The 'vee' windscreen was retained for early production 'E's before being replaced by an improved flat screen. The same sliding canopy and ejection system remained and the armament of six 0.50-calibre M-3 machine guns was unchanged.

F-86F

The most numerous day-fighter version of the Sabre was the F-86F, although this version was also produced to fulfil the fighter-bomber role, a mission at which it had excelled in Korea by 1953. A total of 2,540 F-86Fs were built when production ceased in October 1955 and the last deliveries occurred in December 1956. The F-86F was built at both the Inglewood plant and at the former Curtiss-Wright factory in Columbus, Ohio. This additional assembly location was operated by NAA as the factory in California was at maximum production. The first F-86F-1, 52-2850, flew on 19 March 1952 with J. Pearce at the controls. F-86Fs arrived in Korea within three months of introduction and first equipped the 39th Fighter Squadron of the 51st Fighter Group.

Initial 'F'-model Sabres were assembled with older-model J47 turbo-jets due to engine shortages, but eventually the upgraded General Electric J47-GE-27 jet engine developing 5,910lb of thrust (12% more than the -13) was fitted.

The F-86F had rerouted hydraulic lines to minimise damage to both its normal and alternate systems if hit by enemy fire. The 'F' also featured a revised cockpit layout. Due to the improved thrust to weight ratio the 'F' was nearly 10 miles per hour faster than the 'E' and the time to climb to 30,000ft was about a minute faster. The service ceiling, at 48,000ft, was slightly better. Internal fuel remained the same, at 435 US gallons, but with the F-86F-5, the maximum fuel increased to 835 US gallons from 675 US gallons on the 'E' by adding stronger shackles beneath each wing to handle 200-gallon drop tanks instead of 120-gallon tanks used on the earlier models. This increased tankage added an extra 20 minutes to the F-86F in the combat zone.

The F-86F added a new and easier to maintain radar ranging type A4 gyro computing gun sight and a new manual pip control bomb-aiming device. The AN/AP-30 radar provided range data and automatically locked on to a target. Some 'F's also introduced the '6-3' wing (see p. 123).

Later 'F's introduced an additional wing pylon closer to the fuselage, which allowed two 1,000lb bombs or two 120 US gallon drop tanks to be carried in addition to the usual 200 US gallon drop tanks. This added considerably to the Sabre's fighter-bomber capability. A few F-86Fs carried the very complex Low Altitude Bombing System which was required to deliver a nuclear weapon while allowing the F-86 to escape the blast.

The final 'F' was the F-86F-40. It was first built for the Japanese Air Self Defence Force. Its outer wing was lengthened by a foot on each side and returned to using leading edge slats. The longer wing reduced the stalling speed and allowed slower landing approaches and shorter take-off distances. The result was

an overall improvement over Sabres with the '6-3' wing. The conversion also improved the combat radius and the Sabre's high-altitude manoeuvrability. Because of this, the USAF converted all active-duty Sabres to F-40 specification beginning in March 1955. NAA built 280 F-86F-40 Sabres and Mitsubishi in Japan assembled 300 more.

Many F-86F-40 Sabres could carry AIM-9B Sidewinder missiles. Some 'F's were later modified into radio-controlled target planes and designated QF-86F.

NAA TF-86F

Two F-86F-30s, serial numbers 52-5016 and 53-1228, were modified into two-seat trainers and designated TF-86F. This was to be a potential replacement for the T-33 trainer. Each aircraft had its fuselage lengthened to 42ft 9in. The first TF-86F flew on 14 December 1953. Both were capable of 692mph with a service ceiling of 50,500ft. 52-5016 was lost in a crash early in its career, but 53-1228 served at Edwards Air Force Base as a chase plane for seven years. The TF-86F programme, however, was cancelled in favour of the later F-100F programme.

F-86H

The fighter-bomber version of the Sabre fitted with a General Electric J73 turbo-jet (a substantially more powerful engine than the F-86F's J47-GE-27), deeper fuselage, larger intake duct, greater fuel capacity, larger tail-plane without dihedral, electrically operated flaps, hydraulically operated speed-brakes and controls, stronger landing gear, improved suspension and release mechanism for carrying droppable wing tanks in conjunction with bombs and rockets. Clam-shell-type canopy (similar to that of the F-86D), superior armament (four 20mm M-39 cannons, beginning with the 116th produced) and an improved ejection seat. This version could carry up to 2,000lb of bombs, including, when fitted with the Low Altitude Bombing System (LABS), nuclear weapons. Some 473 were built.

F-86K

NATO version of F-86D, which had 'less classified' avionics fitted.

F-86L

Upgrade conversion of F-86D with new electronics, extended wing-tips and wing leading edges, revised cockpit layout and uprated engine; 981 converted.

NAA FJ Fury

The FJ Fury was a carrier-capable version of the F-86 used by the US Navy and Marine Corps. These aircraft featured folding wings, wider track landing gear and a longer nose landing strut designed to both increase angle of attack upon launch and to absorb the shock of hard landings on an aircraft carrier deck. The windscreen was also modified to give the pilot a better view during approach.

FJ-2: naval day-fighter based on the F-86E, but armed with 20mm cannon. The engine was the General Electric J47-GE-2, a navalised version of the J47-GE-27 from the F-86F. The naval modifications of the FJ-2 had increased weight by about 500kg over the F-86F, but unfortunately had not succeeded in delivering a fully carrier-capable fighter. A decision had already been made to give it to land-based squadrons of the US Marine Corps.

Construction was slowed due to demand for the F-86 in Korea; the FJ-2 was not produced in large numbers until after that conflict had concluded. Only seven aircraft had been delivered by the end of 1953, and it was January 1954 before the first aircraft was delivered to a Marine squadron, VMF-122. The Navy preferred the lighter F9F Cougar due to its superior slow-speed performance for carrier operations, and most of the 200 FJ-2 models built were delivered to the US Marine Corps. The Marines did make several cruises aboard carriers and tried to solve the type's carrier handling problems, but the FJ-2 was never really satisfactory.

FJ-3: a development of the FJ-2 version powered by the Wright J65, a licence-built version of the British Armstrong-Siddeley Sapphire turbo-jet. The Sapphire promised to deliver 28% more thrust than the J47, for little gain in weight. Because of its more powerful engine, the FJ-3 was superior to most models of the F-86, except the F-86H. A total of 538 FJ-3s were built.

All Furys were built in Columbus, Ohio.

CAC Sabre (Australia)

Based on the F-86F airframe but featuring a different engine and built under licence by the Commonwealth Aircraft Corporation (CAC) in Australia, for the Royal Australian Air Force. The CAC Sabres included a 60% fuselage redesign, to accommodate the Rolls-Royce Avon Mk 26 engine, which had roughly 50% more thrust than the J47, as well as 30mm ADEN cannon and AIM-9 Sidewinder missiles. As a consequence of its power plant, the Australian-built Sabres are commonly referred to as the 'Avon Sabre'. CAC manufactured 112 of these aircraft.

Canadair Sabre

The F-86 was also manufactured by Canadair in Canada as the CL-13 Sabre initially to replace its de Havilland Vampires, but it went on to be a major export success with the following production models:

- Sabre Mk 1: one built, prototype F-86A.
- Sabre Mk 2: 350 built, F-86E type, 60 to USAF, 3 to Royal Air Force, 287 to RCAF.
- Sabre Mk 3: one built in Canada, test-bed for the Orenda jet engine.
- Sabre Mk 4: 438 built, production Mk 3 (but retained the General Electric engine), 10 to RCAF, 428 to RAF as Sabre F4.
- Sabre Mk 5: 370 built, F-86F type with Orenda engine, 295 to RCAF, 75 to West German Luftwaffe.
- Sabre Mk 6: 655 built, 390 to RCAF, 225 to West German Luftwaffe, 6 to Colombia and 34 to South Africa.

MAIN DIFFERENCES TABLE

F-86 SERIES ITEM	F-86A	F-86D	F-86E	F-86F
Engine	J47-GE-7 or -13	J47-GE-17 (afterburner)	J47-GE-13	J47-GE-27
Engine Control	Mechanical	Electronic	Mechanical	Mechanical
Automatic Pilot	No	Yes	No	No
Horizontal Tail	Conventional	Single controllable surface	Controllable stabilizer and elevator	Controllable stabilizer and elevator
Aileron & Horizontal Tail Control	Conventional and hydraulic boost	Full-power hydraulic irreversible control	Full-power hydraulic irreversible control	Full-power hydraulic irreversible control
Wing Flap Actuation	Hydraulic or Electric	Electric	Electric	Electric
Aileron & Horizontal Tail Artificial Feel System	No	Yes	Yes	Yes
Armament	Machine guns, bombs, rockets, or chemical tanks	Rockets in fuselage package	Machine guns, bombs, rockets, or chemical tanks	Machine guns, bombs, rockets, or chemical tanks
Gun Charging	Manual (on ground)	No	Manual (on ground) and/or automatic	Manual (on ground) and automatic
Windshield	Curved or "V"	Flat	"V" or flat	Flat
Canopy Ejection Control	Handle on pedestal Right handgrip on seat	Right handgrip on seat	Right handgrip on seat	Right handgrip on seat
Canopy	Sliding	Clamshell	Sliding	Sliding
Oxygen Regulator	A-14	D-1	A-14	A-14 or D-1

Index